Interior Trim

Making, Installing & Finis

WILLIAM P. SPENCE

Sterling Publishing Co., Inc.
New York

DISCLAIMER

Although the information presented in the following pages was provided by a wide range of reliable sources, including material and tool manufacturers, professional and trade associations, and government agencies, it should be noted that the use of tools and materials in home-maintenance activities involves some risk of injury. The reader should always observe the local building codes, the operating instructions of equipment manufacturers, and directions of the companies supplying the materials. The author and publisher assume no liability for the accuracy of the material included.

PHOTO CREDITS

Photos on pages 35, 107, and 145 courtesy of Sherwin-Williams; page 4 courtesy of Kolbe and Kolbe Millwork Co., Inc.; pages 48 and 126 courtesy of Ornamental Moldings; page 71 courtesy of Turnkeyhomes.com; page 94, courtesy of Therma Tru Doors; and page 173 courtesy of Harris Tarkett

Library of Congress Cataloging-in-Publication Data

Spence, William Perkins, 1925-
   Interior trim: making, installing & finishing / William P. Spence.
       p. cm.
   Includes index.
   ISBN 0-8069-9297-2
  1. Trim carpentry. I. Title: Interior Trim: Making, Installing & Finishing. II. Title.
TH5695 .S64 2003
694'.6—dc22

                        2003017540

2 4 6 8 10 9 7 5 3 1

Published by Sterling Publishing Co., Inc.
387 Park Avenue South, New York, NY 10016
© 2004 by William P. Spence
Distributed in Canada by Sterling Publishing
℅ Canadian Manda Group, One Atlantic Avenue, Suite 105
Toronto, Ontario, Canada M6K 3E7
Distributed in Great Britain by Chrysalis Books Group PLC
The Chrysalis Building, Bramley Road, London W10 6SP, England
Distributed in Australia by Capricorn Link (Australia) Pty. Ltd.
P.O. Box 704, Windsor, NSW 2756, Australia

Printed in China
All rights reserved

Sterling ISBN 0-8069-9297-2

Contents

Interior Trim & Woodwork

Interior trim means different things to different people. Basically, it refers to the use of moldings to construct door and window casings[1], chair rails[2], baseboards[3], and cornices[4]. However, it can include other interior woodwork such as wall frames, wainscoting[5], paneling, fireplace mantels and surrounds, columns, and pediments[6] (**1–1**).

The major purpose of installing casing is to provide a decorative feature to the room. In addition, it covers unsightly connections such as those between the drywall (also known as gypsum wallboard) and the door and window frames. On exterior walls, the space between the frames and wall studs is filled with insulation. The casing covers this and helps seal the wall, reducing air leakage into and out of the room.

1–1 A wide range of trim and interior woodwork has been used to produce a warm and very attractive room. The trim, paneling, shelving, fireplace surround, and ceiling beams all blend together, producing a harmonious scene. **Courtesy Architectural Paneling, Inc.**

In addition to covering gaps, the trim, if properly installed, allows the materials involved to expand and contract. For example, the baseboard is nailed to the wall studs and bottom plate. The flooring goes under it and can expand and contract along the wall without showing a crack.

[1] *Casings are the enclosing frames around a door or window.*
[2] *Chair rails are horizontal moldings affixed to a wall equal to the height of the back side of a chair.*
[3] *Baseboard is the molding covering the joint where the wall meets the floor.*
[4] *A cornice is a horizontal molding that crowns an architectural composition.*
[5] *Wainscoting is the lower three or four feet of an interior wall when it is finished differently from the remainder of the wall.*
[6] *Pediments are triangular decorative forms used over doors and windows.*

Baseboard and chair rails protect the wall from damage (**1–2**). Wall frames, wainscoting, and paneling form a major decorative feature of the room (**1–3**). A fireplace will work fine without a mantel or surround, but is much more appealing when the decorative woodwork is installed (**1–4**). The transition between the wall and ceiling can be enhanced with a decorative cornice (**1–5**). Crown molding is the simplest form of a cornice.

Various decorative inlaid and overlaid wood carvings provide a focal point for some feature in the room, such as, for example, placing a carving in the center of the fireplace surround (**1–6**).

Trim can be used to change the character of a room and connecting rooms. For example, a

1–2 Baseboard and chair rails are not only decorative, but protect the wall from damage. The chair rail also provides a place to change the character of the wall covering. Courtesy Mr. and Mrs. James Sazama

1–4 When a fireplace installation includes a wood surround or a mantel, the entire wall is enhanced. This surround is the same material as the paneling.

1–3 Paneling provides a warm, rich wall finish. It will mellow with time. Notice the wide, multilayer cornice at the ceiling.

1–5 This elaborate cornice provides an attractive transition between the wall and ceiling.

1–6 This decorative wood carving is overlaid on a panel.

room with the bare minimum of trim presents a rather stark image (**1–7**). If features such as a small cornice, wall panels with wainscoting, and hardwood trim are used, the appearance is totally different (**1–8**). When horizontal trim such as a chair rail or wainscoting is placed along the

1–7 A typical minimum-trim, paint-grade installation is rather dull.

1–8 Adding crown moldings, using hardwood trim, staining the doors, and adding wainscoting greatly enhance the overall appearance of the room. Courtesy Kolbe and Kolbe Millwork Co., Inc.

walls, it can make a small room seem larger. A low ceiling can be made to appear a bit higher by installing vertically applied moldings such as pilasters[7], paneling with a vertical emphasis (**1–9**), or tall wall frames.

A warm, rustic look may be created by installing wood ceiling beams and adding hardwood crown moldings along the ceiling. This will emphasize the length of the room and add a feeling of warmth and coziness (**1–10**).

The use of interior columns adds a touch of class to any room, especially large rooms with high ceilings. They are finished, so they blend in with the trim and enhance the finished wall (**1–11**).

An outstanding example of the use of interior woodwork to dramatically enhance a large room

[7]Pilasters are tall fluted pillars often with a capital and a base.

1–10 This room has hardwood ceiling beams (top) and crown molding that emphasize the length of the room and create a rustic, cozy atmosphere.
Courtesy Mr. and Mrs. James Sazama

1–9 This wall has tall, narrow wall panels plus a fluted pilaster that help make the ceiling seem higher.

with a high ceiling is shown in **1–12**. The hardwood moldings are assembled into a ceiling beam, terminating in a circular installation in the center of the room.

Another attractive interior trim installation is a **niche**. The niche shown in **1–13** is cast from a polymer plastic and is recessed into the wall. Smaller, surface-mounted niches are also available.

Extending a wide cornice to enclose the mechanism for draperies is a nice way to smoothly tie them into a single visual element (**1–14**).

1–11 This high ceiling foyer uses interior columns to enhance the dramatic entrance. Notice how the walls are lightly tinted and blend smoothly with the white trim and columns. **Courtesy Melton Classics, Inc.**

1–12 This wood ceiling beam and circular wood focal point create a dramatic impression on those who enter the room. **Courtesy White River Hardwoods Woodworks and The Hardwood Council**

1–13 This recessed polymer niche provides an exceptional decorative feature and focuses attention on the art object displayed. **Courtesy Architectural Products by Outwater, LLC (800-835-4400)**

1–14 The extension of a wide cornice to enclose the drapery mechanism is an excellent way to blend these items together. **Courtesy Architectural Products by Outwater, LLC (800-835-4400)**

CHOOSING TRIM STYLES

Large homes built in the early years of the United States placed considerable emphasis on interior trim. Over the years, styles were established that related to homes of that period. Homes built today tend to use a simpler variety of interior moldings. However, as you plan your home, give special consideration to the style and character of the interior trim. It can make a rather plain, dull room take on a totally different character. The trim, plus the selection of appropriate wall finishing materials, can add warmth to the room and reflect a certain style, such as the simple, clean lines of a contemporary home (**1–15**) or the ornate trim from an earlier period (**1–16**).

1–15 This contemporary home features simple lines and minimum emphasis on the interior trim.
Courtesy Kolbe and Kolbe Millwork

1–16 This elaborate cornice uses moldings popular in Early American homes.

The style of trim you select has a great influence upon the final result and your satisfaction. It is worth the time to view the stock styles available, and even consider having some major parts custom-made to your design. If the house is of a particular style, such as Old Victorian or Early American, the moldings used during that period would likely be most suitable. If you have a contemporary house in which trim was typically simple and inconspicuous, a different choice would be made. If you have a new house but want the interior to reflect a certain period, again, select the proper trim.

The local library will most likely have books on early architecture that are a good source of designs. Some molding manufacturers will have traditional molding designs, as well as many very

special trim products available (**1–17**). Your local building-materials dealer will have catalogs from which to order these special trim materials.

SELECTING MOLDING

The first decision is whether the molding will be painted, finished clear, or stained. When remodeling an old classical house, it is necessary to match the existing trim; this may turn out to be a costly proposition. It may require custom-milled profiles and an expensive wood such as cherry or walnut.

If building a new house, you have a lot more freedom. Your choices will be influenced by the appearance desired and the cost. If you use stock moldings, you have a choice of profiles and sizes, and the option of combining several to produce a more detailed molding. You will be limited on a choice of materials.

The best overall appearance is produced if the trim in a room all represents one style. For example, mixing a Colonial trim and something with a Greek or Roman style does not produce the best results.

If all the trim in the room is new, check the molding thickness before buying any to be certain that where pieces butt together they will be flush on the surface. If it is a remodeling job and only some of the trim is being replaced, take a piece of the old trim with you when you select

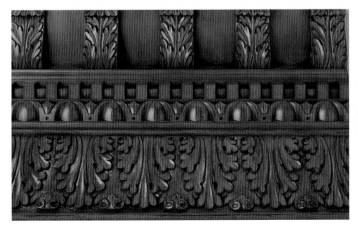

1–17 This deeply sculpted crown molding is an example of the high-quality classical products available. Courtesy Architectural Paneling, Inc.

the new material. A typical problem can occur where a baseboard butts the casing on a door. Using old trim as a reference will ensure that the new trim has the same profile as what is already in the room.

Before you actually buy the molding, be certain how it will be finished. If it is to be painted, the choice is easy. If a clear or stained finish is desired, the particular species of wood must be chosen—and do not forget to get the cost. If using polymer molding, get the manufacturer's finishing instructions. (Refer to Plastic Moldings on page 18 for more information.)

CHOOSING A WOOD SPECIES

While wood molding is available in a number of species, frequently the choice at the local building-supply dealer will be between a softwood and oak. Manufacturers also produce molding in various hardwoods by special order.

Experienced finish carpenters know which woods are the best to work when grain, ease of cutting, sanding, shaping, and nailing are considered. Most softwoods are easy to process and nail. Cutting the profile on hardwoods is a bit more difficult, but not a major consideration. However, hardwoods must often be drilled for nailing. Open-grain woods will have to have a coat of filler applied to fill the grain before it can be painted or stained.

Hardwoods wear better than softwoods and resist bumps and abrasion. If the moldings are wide or in a damp location, they will expand when the humidity is high. Some woods are more stable than others and less affected by moisture. Moldings made from polymers are not affected by moisture and serve well in moist locations.

Availability is another factor to consider. The local building-supply dealer will most likely have a limited supply of stock wood moldings. Ascertain the possibility of getting

more detailed profiles of hardwoods instead. These will have to be ordered. What is the chance that the dealer can actually secure these by the time you need to install them? What will they cost?

CHOOSING STOCK OR CUSTOM-MADE MOLDINGS

When choosing molding, you have to decide whether the stock moldings available will produce the image you desire, or if the moldings must be custom-made to conform to the specifications of your drawing. In some cases, a finish carpenter will have the equipment and knowledge to custom-make the molding for you. Local woodworking shops can also produce moldings.

If the stock moldings available appear satisfactory, profiles can be combined to produce a larger, more detailed molding (**1–18**). You can also combine a custom molding that has a special feature with one or more stock moldings.

Profiles for a few of the wood stock moldings commonly available are shown in **1–19**.

Manufacturers produce a wide range of wood moldings that provide that special look for a room and are used to trim rooms in the classic style of a past era (**1–20**). They are made from select hardwoods such as walnut, mahogany, cherry, poplar, and oak. Hardwood moldings require a surface free or defects and perfectly fitting joints.

Wood trim that is to be painted must be made from a wood that machines and sands to a smooth finish. Pine is probably the most commonly used wood. Both white and yellow pine are available in various sections of the United States. Poplar, a hardwood, is also favored because it is inexpensive and wears well. In western states, fir is often used for paint-grade, natural-finish, and stain-grade moldings. Paint-grade moldings may have

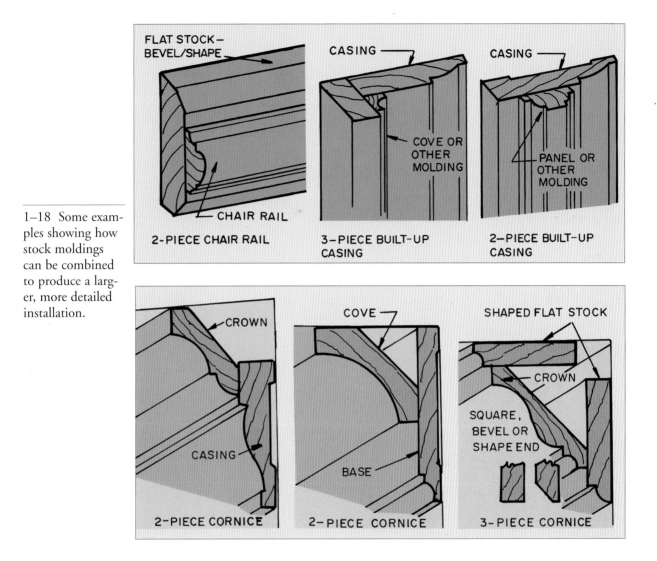

1–18 Some examples showing how stock moldings can be combined to produce a larger, more detailed installation.

minor blemishes requiring the painter to fill the blemishes, set and caulk the nails, and caulk poorly fitting joints.

Paint-grade pine moldings made of finger-jointed short pieces are less costly than defect-free, single pieces. When sanded carefully and painted, the joints are seldom seen. Finger-jointed wood moldings are also available covered with a solid-wood veneer or a wood-grained vinyl sheet. These are cut and installed in the same way as solid-wood moldings. However, care must be taken so that the veneer or vinyl covering is not damaged (1–21).

Hardwood molding is often available in oak, poplar, cherry, walnut, and mahogany. However, most molding manufacturers will machine out molding in any type of hardwood you choose. These are beautiful, but are more expensive than paint-grade moldings. In addition, the manufacturer will produce custom profiles in any of these woods.

If you are going to machine the molding yourself, a good grade of lumber is necessary. It must be straight, kiln-dried for interior use, and free of knots, pinholes, splits, and checks.

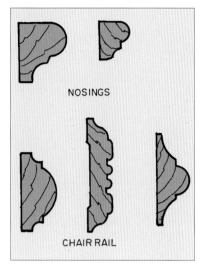

NOSINGS

CHAIR RAIL

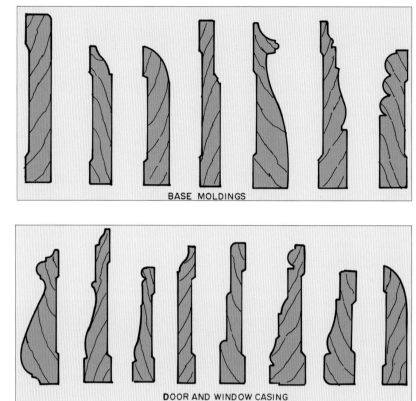

BASE MOLDINGS

DOOR AND WINDOW CASING

1–19 A few profiles available on the most commonly used interior moldings.

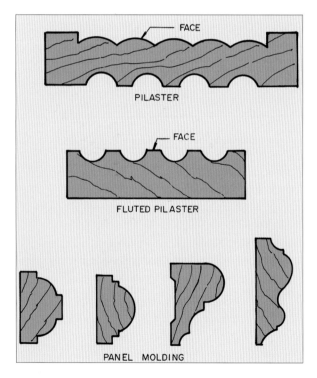

FACE

PILASTER

FACE

FLUTED PILASTER

PANEL MOLDING

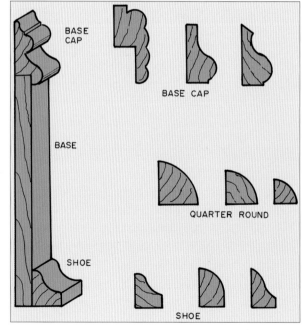

BASE CAP

BASE CAP

BASE

QUARTER ROUND

SHOE

SHOE

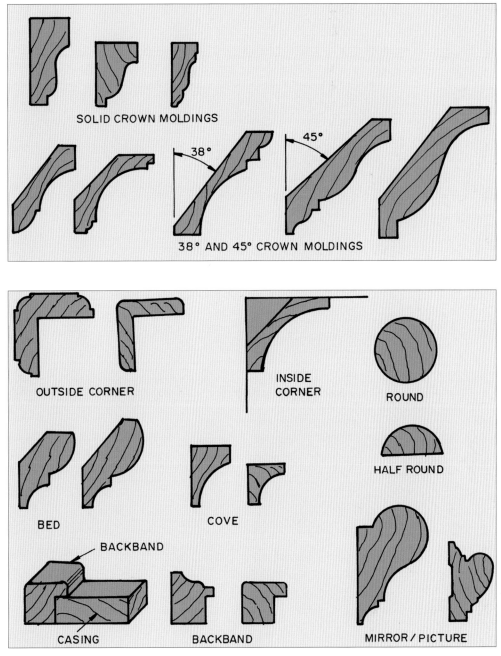

SOLID CROWN MOLDINGS

38°

45°

38° AND 45° CROWN MOLDINGS

OUTSIDE CORNER

INSIDE CORNER

ROUND

BED

COVE

HALF ROUND

BACKBAND

CASING

BACKBAND

MIRROR / PICTURE

1–19 continued.

1–20 These are examples of the beautiful hand-carved moldings available. They are carved from select hardwoods. Moldings such as these enhance the overall ambience of a room. Courtesy Architectural Paneling, Inc.

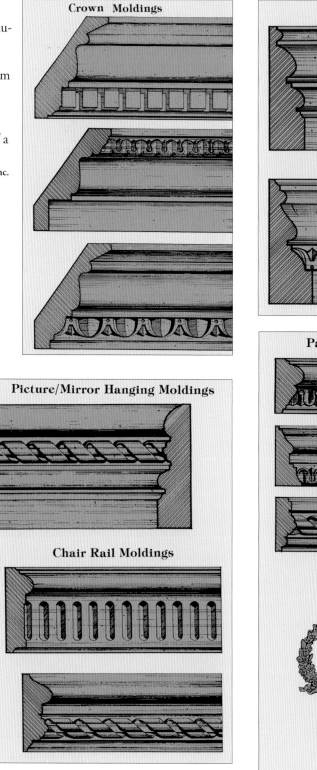

Crown Moldings

Picture/Mirror Hanging Moldings

Chair Rail Moldings

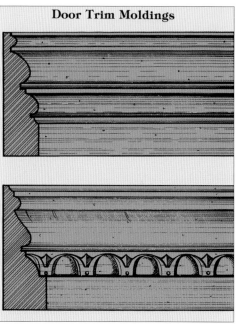

Door Trim Moldings

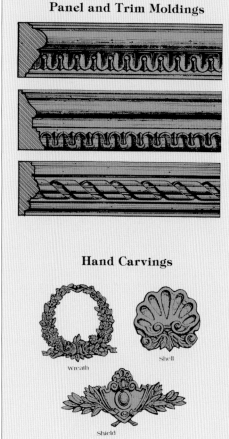

Panel and Trim Moldings

Hand Carvings

Wreath

Shell

Shield

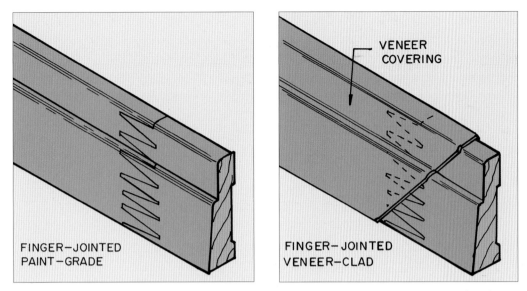

1–21 Finger-jointed moldings are used when the molding is to be painted or covered with a veneer.

WOOD TRIM GRADES

Trim grades for wood products indicate the quality of the material. While the terms used to identify each grade may vary across the country, basically they fall into two broad categories—**paint-grade** and **stain-grade**. Each grade is then divided into several subcategories. Stain-grade refers to high-quality products generally reserved for interior trim. They are more expensive than paint-grade products. If you can find a top-quality paint-grade product, it might be stained for use in some rooms; this will reduce cost. Basically, for top-quality jobs where clear or stained trim is desired, it is best to use stain-grade trim.

AWI STANDARDS

The Architectural Woodwork Institute (AWI), 1952 Isaac Newton Square, Reston, VA 20190, is a not-for-profit organization of manufacturers of architectural woodwork located in the United States and Canada. Its publication *Architectural*

Woodwork Quality Standards provides information used to specify architectural woodwork. This enables bidders on architectural woodwork to have the opportunity to compete on an equal basis to perform work of equal quality.

Architectural woodwork is used primarily on very fine projects. While it is most commonly used on commercial buildings, high-quality residential work is often bid based on these standards.

The three AWI grades are premium, custom, and economy. **Premium grade** is designated when the highest degree of control over the quality of the materials, workmanship, and installation is required. It is reserved for special projects such as large, curved stairways or rooms with hardwood panels and elaborate cornices. The wood is matched for color and grain.

Custom grade indicates a well-defined degree of control over quality of workmanship, materials, and installation. It is the level expected in high-quality custom-built homes. The vast majority of architectural woodwork falls in this category.

Economy grade represents the minimum acceptable quality of workmanship, materials, and installation within the scope of AWI standards. It is what is typically used in the average residential project, where attempts are made to keep the cost low.

BUYING & STORING WOOD MOLDINGS

Frequently, moldings ordered from the local building-supply dealer will have pieces that are in worse condition than expected. They are often stored in outbuildings where there is no moisture control. They get banged around as people look through the supply, put them on the truck, and deliver them to the site. If you go to the dealer and are permitted to select the pieces yourself, check to see that they are not warped or twisted.

Bring the moldings into the house only after the drywall has been taped and dried. Place them flat on the floor and put sticks between them so air can circulate. Leave them there a week or so to let them adjust to the moisture level of the air in the house. The room should be closed and, if possible, the heating/cooling system operational to control moisture in the air.

Since the actual size of the moldings may vary somewhat from their given size, it is wise to try to buy all the material at once to ensure that the moldings you get were all run at the same time. That way, they will most likely be the same size.

PLASTIC MOLDINGS

Interior trim is available that has been made from polymers. Typically, fiberglass-reinforced polyester and extruded polystyrene are used. They are available with a factory-applied prime coating ready for painting, or prefinished with an artificial wood-grain pattern. If painting these moldings, use a good-quality oil- or latex-based paint. Never coat plastic moldings with a lacquer-based product.

Plastic moldings are available in complex, detailed profiles representing many of the classical wood moldings found in early buildings (**1–22**). Since they are a single piece, they are easier to install than multipiece wood moldings of the same design. Rigid polymer moldings are lightweight and can be cut with regular woodworking tools. They are bonded to the wall with an adhesive supplied by the manufacturer. Sometimes, they need to be held with a few nails until the adhesive cures. While plastic moldings are sometimes more expensive than wood, the savings in installation time are considerable.

1–22 Polymer interior moldings are available in very detailed, complex forms. They consist of a single piece and are easier to install than complex wood moldings made from several separate pieces. **Courtesy Architectural Products by Outwater, LLC (800-835-4400)**

FLEXIBLE PLASTIC MOLDINGS

Some types of plastic molding are very flexible and may be bent to follow specific contours (**1–23**). They are easily formed around arched openings, large-diameter openings, and columns. They can be easily installed as baseboards, crowns, and casings on walls that have rounded surfaces.

Some types are more flexible than others (**1–24**), so take this into consideration when choosing a molding. Plastic moldings are hard enough to be safely nailed with pneumatic nailers. They can be cut with a saw or a knife. Some are installed with a manufacturer-supplied adhe-

1–23 Flexible plastic moldings are easy to bend and conform to curved applications. **Courtesy Flex Trim, Inc.**

advance notice so the desired profiles can be ordered from the plastic-molding company. Moldings are available in lengths from 6 to 16 feet, depending upon the profile and manufacturer.

MEDIUM-DENSITY FIBERBOARD MOLDINGS

Medium-density fiberboard (MDF) (**1–25**) is made from a mixture of wood fibers, wax, and a resin binder. It is compressed into the molding profile under high pressure and is available in a range of molding types and profiles. The finished molding is factory-primed, making it easy to paint and produce a nice finished appearance.

MDF is very hard and should be cut with carbide-tipped saws. When being cut, it produces a fine sawdust, so be sure to wear a respirator and protect your eyes. Since it is so hard, it cannot be hand-nailed. Use a pneumatic nailer or screws to secure the molding. When nailed, it tends to push up around the nail head, and this "dimple" has to be chiseled or sanded smooth before the hole is caulked (**1–26**).

1–24 Some flexible plastic moldings can be bent on curves with a short radius. **Courtesy Flex Trim, Inc.**

sive. They are available in stain or paint grades, and with an embossed wood grain.

BUYING PLASTIC MOLDINGS

Plastic moldings are stocked by the local building-supply dealer. As is true with wood, the sizes and profiles available are extensive and not every type of molding will be in stock. Give your dealer

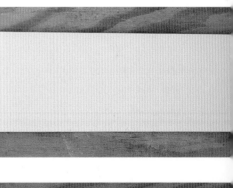

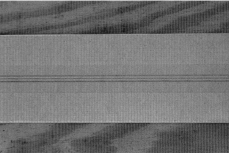

1–25 Factory-primed, medium-density fiberboard molding. The back view shows the brown, unprimed surface.

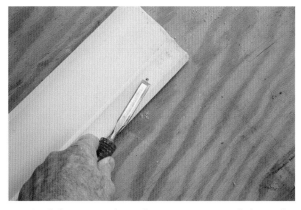

1–26 If the material mushrooms up around the nail hole, trim it flush with a sharp chisel or sand it lightly with a power sander.

OTHER INTERIOR WOODWORK

Pilasters are tall, fluted pillars that often have a capital and a base. When used as interior trim, they serve as a decorative feature and do not support a load. They are typically rectangular or semicircular and are used to simulate pillars on door openings and fireplace surrounds (**1–27**). They are available in wood and plastic.

Columns used as interior trim are for decorative purposes and do not support a load (**1–27**). They are available in round and semicircular styles.

Most designs available were derived from Greek and Roman architecture. These early columns were designed in a special sequence and in a proper proportion to produce a product that is pleasing to the eye. Basically, they consist of a **plinth**, **pedestal**, **base**, **shaft**, and **capital** (**1–28**). They are available in wood and high-density polymer.

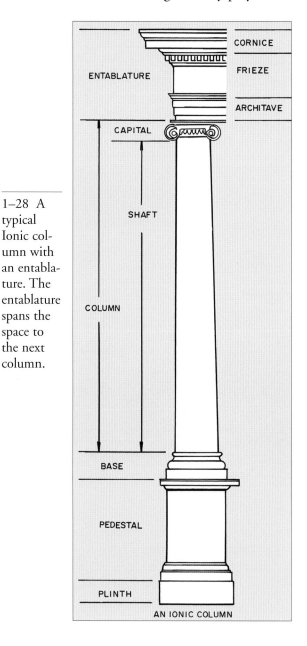

1–28 A typical Ionic column with an entablature. The entablature spans the space to the next column.

1–27 These doors have pilasters on the wall on each side. An architrave extends into the room supported by columns that join these elements together into a unified architectural feature. An architrave is a beam that spans from column to column, resting directly upon their capitals. **Courtesy Architectural Products by Outwater, LLC (800-835-4400)**

Another approach is shown in **1–29**. Here a rectangular column that is only half the thickness of a full column has been installed alongside a door with a pediment. The cornice has been installed around the top of the column.

An **architrave** is a beam that runs from column to column, resting upon their capitals, which are the topmost decorative members on columns or pilasters. The architrave is the lowest member of the entablature.

The **frieze** and **cornice** are above the architrave. Together they form the **entablature** (**1–30**). A **pediment** is built upon the entablature, completing the dominant elevation (**1–30** and **1–31**).

1–29 This installation uses a rectangular column against the wall with the cornice wrapped around it. The column stands against a panel wall beside a door with a pediment. **Courtesy Architectural Products by Outwater, LLC (800-835-4400)**

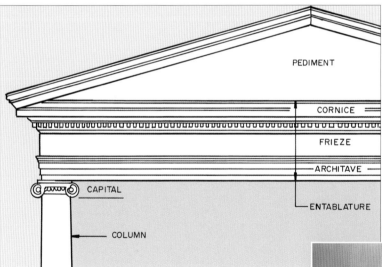

1–30 The cornice, frieze, and architrave form the entablature. The pediment is placed upon the entablature.

1–31 This door has pilaster casings on each side and a pediment crowning the opening. These features are made from molded high-density polyurethane. **Courtesy Architectural Products by Outwater, LLC (800-835-4400)**

A pediment is a triangular facade that has a horizontal cornice and raking cornices. It is used to crown an opening. When used as interior trim, it typically spans a window or door opening.

Fireplace surrounds are another frequently used type of interior woodwork. They may consist of just a mantelshelf mounted above the firebox opening, or a complete surround composed of wood trim on the sides of the opening (**1–32**).

Another interior finishing product is **paneling** used for constructing wainscoting or for covering an entire wall (**1–33**). Manufacturers have wood paneling available in various profiles and wood species. Paneling establishes the atmosphere in the room. The type of wood and the finish make a considerable difference (**1–34**).

1–33 The massive fireplace surround, blended with rich dark paneling, extensive cornice construction, and a broken pediment over the mirror, provides a warm, secluded atmosphere. Courtesy Architectural Paneling, Inc.

1–32 This fireplace surround has fluted side members resting on a plinth and a detailed mantel spanning the opening. Courtesy Heatilator

1–34 The light finish on this paneled door casing and cabinetry provides a brighter atmosphere that is considerably different from that in **1-33**. Courtesy Architectural Paneling, Inc.

ESTIMATING TRIM QUANTITIES

The quantity of base, shoe, ceiling, and chair-rail moldings is determined by finding the linear feet of the perimeter of the room plus any related areas, such as closets, and adding 10 percent for waste. Do not deduct anything for door openings. Moldings are typically available in lengths ranging from 6 to 16 feet. Buy the longest pieces possible.

The quantities of door and window casings needed can be figured by calculating the linear feet around each opening and adding 20 percent for waste. It is possible to buy these casings in precut kits that are ready to install.

Wall-frame molding quantities can be determined by calculating the linear feet around the perimeter of each molding frame and adding 10 percent for waste (**1–35**).

In order to avoid having to splice trim pieces around the panel, you can measure the length of each piece and see how many you can cut from a standard length of molding. Then order the best lengths for your job. Be certain to add at least two inches to the length of each piece so you have stock available to cut the miters. In the example shown in **1–36**, 434 inches are required, including the waste factor, so four 10-foot pieces (totaling 480 inches) should be adequate.

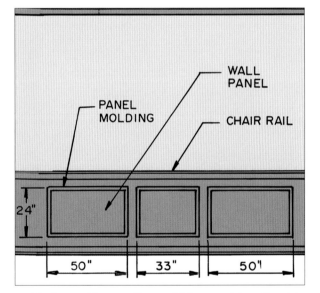

1–35 A panel layout, from which quantities of moldings may be estimated.

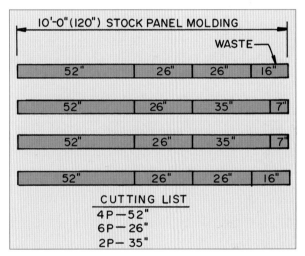

1–36 One possible cutting plan for the panel molding layout shown in **1-35**.

MAKING YOUR OWN MOLDINGS

Begin by choosing the best-quality wood in the species to be used. It should be straight, kiln-dried for interior use (usually with a moisture content of 6 to 8 percent), and as free from pinholes, splits, and checks as possible. Remember, minor defects in paint-grade trim can be filled before painting. This can be done with hardwoods as well, but it is difficult to get the filler to match the final color of the stain. See Chapter 9 for information on finishing.

The main advantage to making your own moldings is that you can form profiles that are not available in stock moldings. It is not, however, usually a way to reduce costs. Molding-grade lumber costs more than the usual lumber available at the lumberyard, and it will probably have to be ordered or purchased from a woodworking shop that makes quality millwork.

These special moldings can be made on the job or in your shop. Typically, they are made with a router, a shaper, a molding head on a table saw, and a molder/planer.

MAKING MOLDINGS WITH A ROUTER

If you frequently use a router to make moldings, get a heavy-duty model. Routers rated at 1½ to 1¾ horsepower have plenty of power to do the job (**1–37**). Be certain to follow the manufacturer's recommendations. Typically, these heavy-duty routers will require bits with a ½-inch shank. Smaller-diameter shanks may bend and vibrate or even break under consistently heavy loads (**1-38**).

Before attempting to make moldings with a router, refer to the information in Chapter 3 concerning router use.

ROUTER BITS

To create a wide selection of profiles, it is necessary to have a good variety of bits. Some commonly available bits are shown in **1–39** and

1–37 This 1¾-horsepower router runs at 25,000 RPM and handles heavy work. **Courtesy Robert Bosch Tool Corporation**

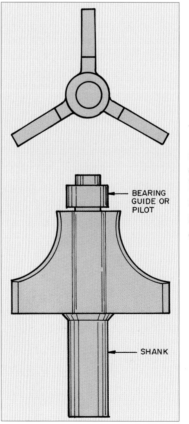

1–38 A typical router bit. This example has a bearing guide that rotates as the bit is moved along the edge of the stock.

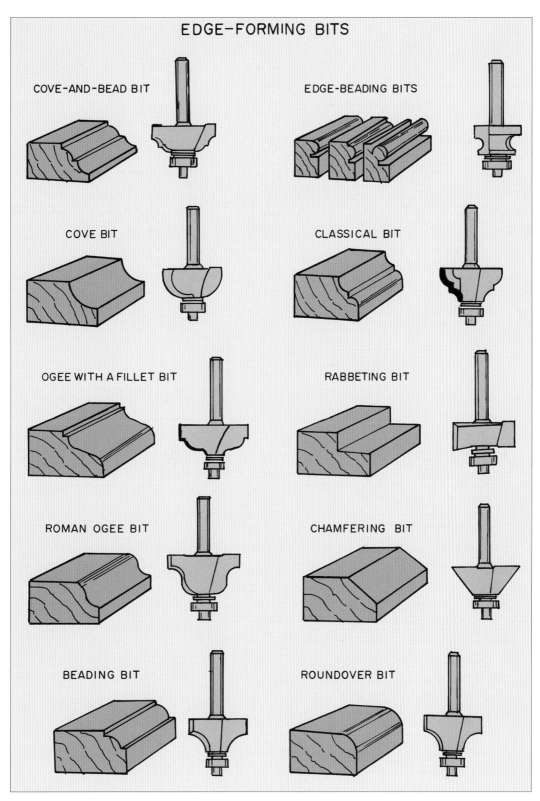

EDGE–FORMING BITS

COVE-AND-BEAD BIT

EDGE-BEADING BITS

COVE BIT

CLASSICAL BIT

OGEE WITH A FILLET BIT

RABBETING BIT

ROMAN OGEE BIT

CHAMFERING BIT

BEADING BIT

ROUNDOVER BIT

1–39 Some of the router-bit profiles available for shaping the edges of stock.

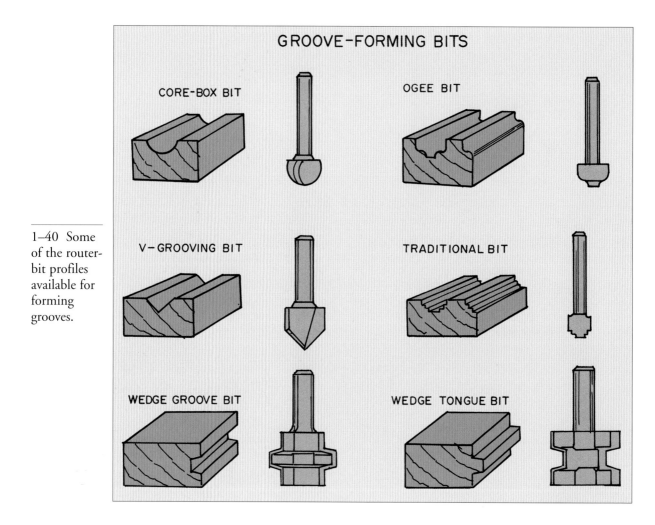

GROOVE-FORMING BITS

CORE-BOX BIT

OGEE BIT

V-GROOVING BIT

TRADITIONAL BIT

WEDGE GROOVE BIT

WEDGE TONGUE BIT

1–40 Some of the router-bit profiles available for forming grooves.

1–40. These are made from high-speed steel or tungsten carbide. Carbide bits are more expensive, but last several times longer than those made from high-speed steel. If you will be making a lot of interior molding, carbide is the better choice. High-speed steel bits can have their profiles changed by grinding, so if you want to alter a profile, they could be the best choice.

Bits used to shape the edges of stock will have a pilot or rolling bearing guide on the shank below the cutters. This rides on the edge of the stock, controlling the depth of the cut (**1–41**). Since the bearing rolls as the cutter moves along the stock, the edge does not get burned. Sometimes, if it is moved too slowly, the pilot bearing will burn the edge of the stock.

1–41 The bearing guide rides along the edge of the stock.

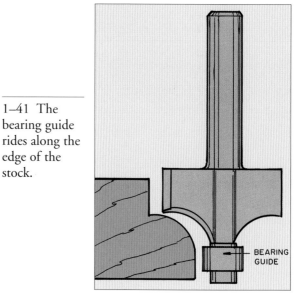

BEARING GUIDE

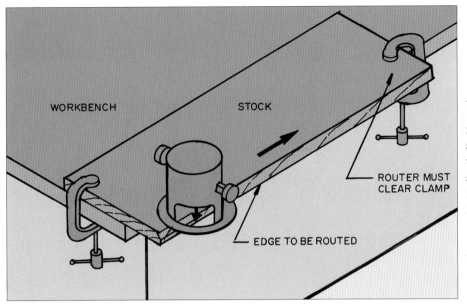

WORKBENCH STOCK

ROUTER MUST
CLEAR CLAMP

EDGE TO BE ROUTED

1–42 Stock to be routed freehand should be securely clamped to a workbench or other stable surface. The clamps must not block the path of the router.

Bits with a rolling bearing are used when the router is used freehand on stock clamped to a workbench. It is recommended to use pilot-bearing guides when the router is mounted on a router table. The fence on the table takes some of the pressure off the pilot, so it is less likely to burn the stock. It also helps to keep the cutters free of resin that may build up from woods such as pine. Lubricate the bearing guides occasionally with a light machine oil so they rotate freely.

ROUTING STOCK FREEHAND

After you choose a bit, install it in the chuck as recommended by the manufacturer. Be certain to remove the router plug from the source of electricity before inserting the bit. After inserting the bit, check to be sure it is secure.

When routing wide stock, clamp it to a workbench as shown in **1–42**. The clamps must be placed so the router clears them. If you are routing narrow strips, mount the router on a router table or use a wood shaper. Now set the depth of cut. The profile can be changed by adjusting how far the bit extends below the router base (**1–43**). The adjustment is made by raising or lowering the router motor. To check the profile,

make a short cut on a piece of scrap stock of the same thickness as the stock to be shaped.

Operating the portable router is relatively easy, but there are some procedures that must be

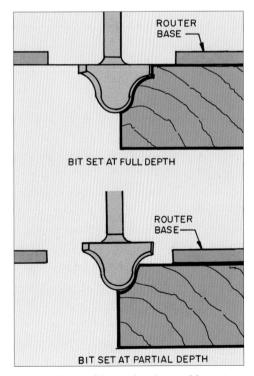

ROUTER BASE

BIT SET AT FULL DEPTH

ROUTER BASE

BIT SET AT PARTIAL DEPTH

1–43 The profile can be changed by raising or lowering the bit.

followed. Wear eye protection, because chips will be flying. Position the base down on the stock with the bit clear of it. Arrange the power cord so it is out of the path of the cut. Hold the router by both handles. Check to be certain that the bit is clear of the stock and turn on the power.

Let the bit reach full speed and then move the router into the stock (**1–44**). Once the pilot or rolling bearing guide touches the stock, immediately move the router along the edge. With the bit rotating clockwise, move the router from left to right (**1–45**). If you are routing end grain, the edge at the end of the pass will often splinter a little. If the end grain is to be shaped, do that first (**1–46**). If there is any splintering on the edge, it is usually small enough that it will be cut away when the edge grain side is routed. If not, try clamping a piece of scrap on the edge of the stock and routing on it.

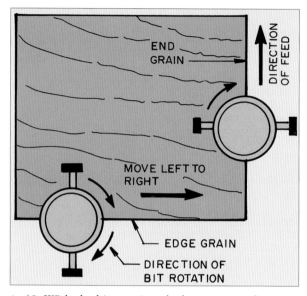

1–45 With the bit rotating clockwise, move the router from left to right.

1–44 After the bit reaches full speed, move the router along the edge letting the bearing guide ride on it. Hold the router by both handles. Courtesy Robert Bosch Tool Corporation

Always keep the router moving. If it stops during a cut, it may burn the edge of the wood. Feed at a constant speed at which the cutter seems to cut easily without undue effort. If you cut too fast, you will get a rough, bumpy surface.

If a large amount of material is to be removed, make the cut in several passes. Set the bit to take a partial cut and set it a little deeper as additional passes are made, until the desired profile has been cut. When cutting several pieces, cut all the pieces at a given depth before resetting the bit to take a deeper cut.

Once the cut has been completed, slide the router away from the stock (but with the base still resting upon it) and turn off the power. Keep it on its base until the bit has stopped turning.

If the molding is to be narrow, you can shape the edge of a wider board and then trim off the width wanted on a table saw, as shown in **1–47**. Use a fine-tooth blade to get a smooth cut. Leave a little extra width so the sawed edge can be smoothed. A jointer is a good machine to use to smooth the sawed edge.

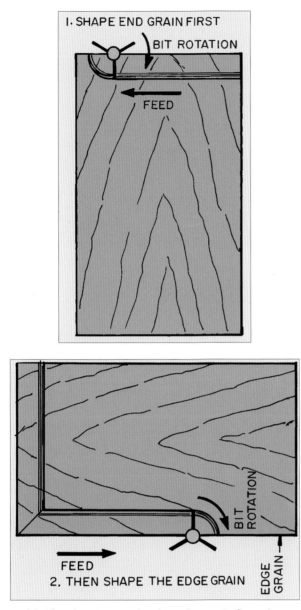

1–46 If end grain is to be shaped, rout it first, then the edge that intersects it.

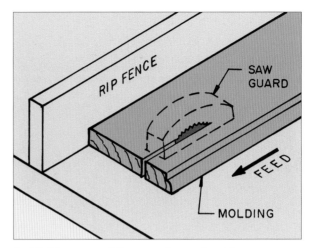

1–47 Narrow moldings can be made with a router by shaping the edge of a wide board and ripping off the width of molding needed.

Another way to run narrow molding is to machine the stock to the desired width and shape it on a wood shaper or a router mounted on a router table.

SHAPING MOLDINGS ON A ROUTER TABLE OR WOOD SHAPER

A router can be mounted under a router table, with the bit extending above the tabletop (**1–48**). This makes it operate much the same as a wood shaper (**1–49**).

A wood shaper (**1–50**) is a much more powerful machine than a router, and is the preferred machine if a lot of molding is to be run. While shapers and routers are both operated in the same basic manner, the router has lower horsepower and small bits, and cannot cut a deep molding in a single pass. It will usually take several passes to make almost any cut with a router. A shaper has higher horsepower and larger, stronger cutters (**1–51**). It can make deeper cuts than the router. Simple profiles can be usually cut in a single pass.

Begin by installing the bit in the router or cutter on the shaper spindle as directed by the manufacturer (**1–52**). Raise or lower the bit or cutter

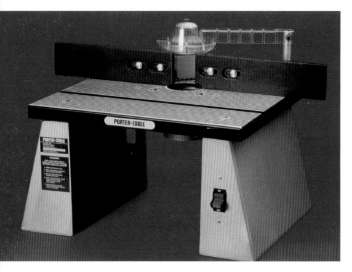

1–48 A router is mounted below the router table with the bit sticking up above the table. **Courtesy Porter Cable Corporation**

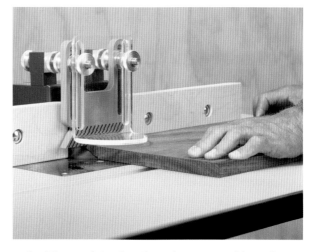

1–49 The stock is placed on the router table and moved past the rotating router bit. **Courtesy Robert Bosch Tool Corporation**

1–50 A wood shaper has a heavy-duty molding cutter above the top of the table. It is suited for heavy-duty shaping work. **Courtesy Delta International Machinery Corporation**

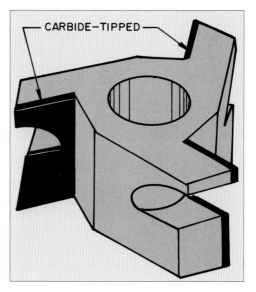

1–51 A typical shaper cutter.

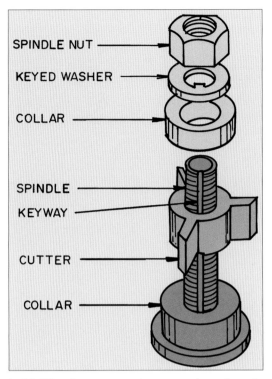

1–52 The shaper cutter is mounted on the spindle. This is a typical assembly.

depth of cut (**1–55**). Safety and operating instructions are included in Chapter 3.

You can get more detailed profiles by using several different bits or cutters. In **1–56**, a cove cut was made on the top of the stock and a multiple bead cut was made on the lower portion.

It is a good practice to run the profile on a piece of scrap stock to verify it is what you want. Then run all the moldings with this setting before you change the setup for some other profile.

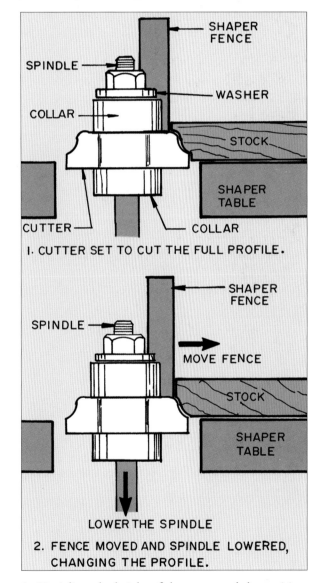

1–53 Adjust the height of the cutter and the position of the fence to get the desired profile.

until the desired profile is above the table (**1–53**). Adjust the fence until the required portion of the bit or cutter extends clear of it. Be certain the table and bit or cutter adjustment screws are tight.

Start the motor and let it get to full speed. Put on safety glasses. Place the stock against the infeed fence and move it to the left into the cutter (if the cutter is rotating counterclockwise). Move it across the table with a uniform speed, allowing the bit or cutter to make the cut without overloading the motor (**1–54**). Move the stock across the cutter and off the outfeed fence, keeping your hands clear of the cutter at all times.

Notice that router bits rotate counterclockwise, so the stock is fed from left to right into the bit. Shapers may have controls to allow the cutter to rotate either way. Always feed the stock into the cutter as shown in **1–55**. Remember to install the cutter so the cutting edge rotates toward the direction of feed. Move the infeed fence to adjust the

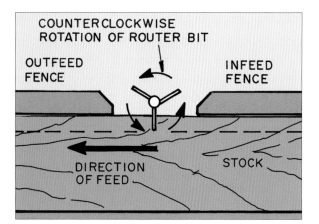

1–54 The router rotates the bit counterclockwise when it is mounted on a router table, so move the stock from right to left.

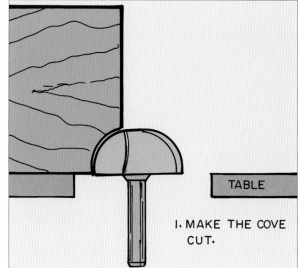

1. MAKE THE COVE CUT.

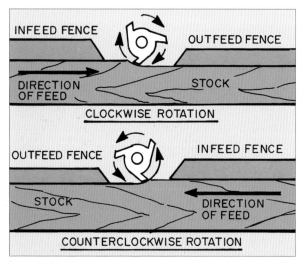

1–55 Shapers can rotate the cutter clockwise or counterclockwise. Feed the stock into the cutting edge of the cutter. Mount the cutter so the cutting edge rotates toward the direction of feed.

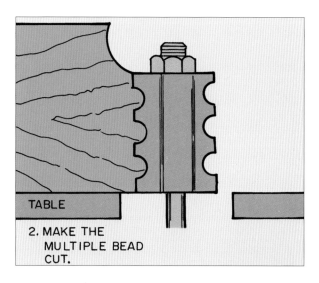

2. MAKE THE MULTIPLE BEAD CUT.

1–56 A profile can be cut by using several different router bits or shaper cutters.

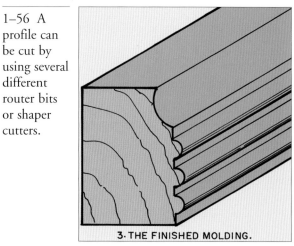

3. THE FINISHED MOLDING.

MAKING MOLDINGS WITH A MOLDING HEAD

A molding head consists of a solid metal disc that has slots machined into it in which the molding knives are secured (**1–57**). Single-, double-, and three-knife models are available. The three-knife models are best. The knives have various profiles ground into them. They are slipped into the slots in the round head and secured with setscrews. A three-knife model will make a smoother cut and be a bit safer than the others.

The molding head is mounted on the arbor of a table or radial arm saw, replacing the saw blade. A wide throat piece provides an opening in the saw table.

Molding heads are very dangerous, and other means of shaping stock should be considered. Since the knives are loose and attached to the head as needed, the danger of throwing a knife as the head rotates is always a possibility. Router bits and shaper cutters on small shapers are single, cast units that run less risk of flying off.

When using the table saw or radial arm saw for any purpose, first refer to the information in Chapter 3 on table saw and radial arm saw use.

MAKING MOLDINGS WITH A PLANER/MOLDER

Those who do a lot of custom molding, or prefer to make their own, often use a planer/molder (**1–58**). This machine planes the surface of

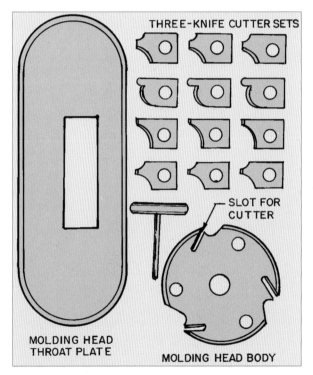

1–57 A molding head is a solid metal disc that has slots machined into it in which the molding knives are secured. The knives are installed in the slots as directed by the manufacturer. Molding heads are very dangerous, and other means of shaping stock should be considered.

1–58 This variable-feed planer will machine stock to the desired finished thickness. In addition, it rips stock to width, sands stock, and cuts molding profiles. Courtesy Woodmaster Tools, Inc.

stock (**1–59**), sands the surface (**1–60**), saws stock to the width desired for the molding (**1–61**), and machines multiple pieces with a molding profile (**1–62**). The speed of feed can be varied to suit the type of work being performed and the characteristics of the species of wood being run. It has two motors. One drives the feed system and the other powers the section performing the planing, sanding, cutting, and molding.

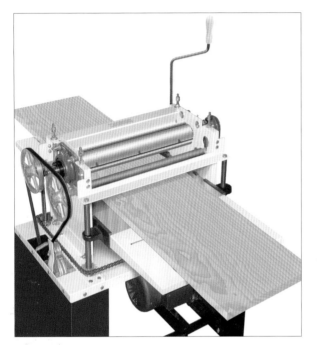

1–60 After the stock has been planed, the sanding head will provide the desired final finish. Courtesy Woodmaster Tools, Inc.

1–59 The planer/molder planes the stock to the desired thickness. Courtesy Woodmaster Tools, Inc.

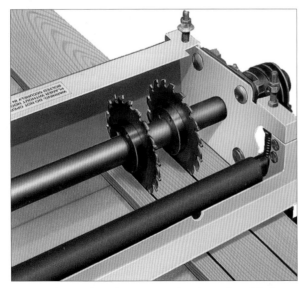

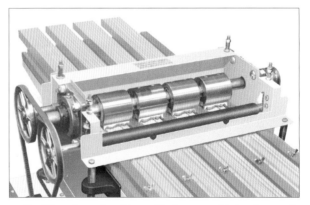

1–62 The planer/molder will shape multiple pieces of stock to the desired profile. Courtesy Woodmaster Tools, Inc.

1–61 The stock is then cut to the desired width for the molding to be run. Courtesy Woodmaster Tools, Inc.

INTERIOR TRIM: MAKING, INSTALLING & FINISHING

CHAPTER 2

Hand Tools

The installation of interior trim does not require the use of many hand tools, but those you do use should be of high quality. When selecting tools for installing interior trim, consider the quality of the materials used and the designs. Poor-quality tools will not produce the results you want. They are a bad investment. Cutting tools, such as planes and chisels, must have high-quality-steel cutting blades and strong handles. They should stay sharp for long periods of time and hold an edge after resharpening. Scraping tools, such as files and hand scrapers, should hold an edge and be easy to clean.

A good method for storing tools is necessary to prevent them from being damaged. Plane blades and the cutting edges of chisels need protection. Saws are especially vulnerable to unexpected damage. Boring tools, such as auger bits, require special attention as well.

If you have a home workshop, wall panels make a good storage area (**2–1**). Finish carpenters also need some type of portable tool chest to keep tools safe on the job (**2–2**). Each tool should have a special place in the tool chest or be kept in a chest specifically designed for it by the manufacturer. Boring tools, such as auger bits, require special protection of the cutting edge and screw tip. Twist drill holders keep twist drills organized and protected (**2–3**).

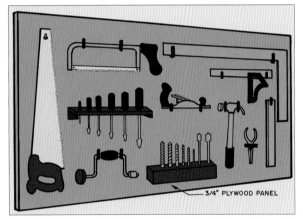

2–1 A tool panel helps keep tools organized and protected.

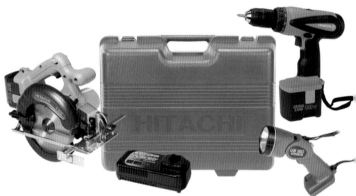

2–2 Various plastic storage boxes help keep tools organized and protected. **Courtesy Hitachi Power Tools**

2–3 Edge tools (left) need protection when stored. These auger bits have protective plastic covers, while the twist drills (right) are kept in a plastic case.

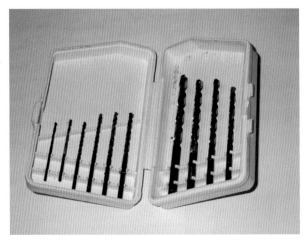

CUTTING TOOLS

The most commonly used cutting tools are shown in **2–4** and **2–5**. A 9¾-inch **smoothing plane** or a 14-inch **jack plane** is used to smooth edges and surfaces of long boards, and even cut a small chamfer or bevel along an edge. A 7-inch **block plane** has the plane iron on a low angle and is used for end- and cross-grain planing. A 3½-inch **trimming plane** is good for light trimming and close work. **Wood chisels** can be used for many trimming jobs, and for cutting joints. Notice that they are stored in a protective plastic case. The familiar **utility knife** will cut just about anything. Remember to replace the blade when it becomes dull. Replacement blades are stored inside the handle.

A

B

C

D

E

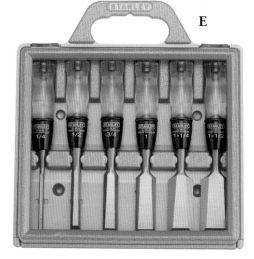

F

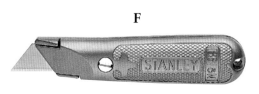

2–4 Frequently used cutting tools include: A, smoothing plane; B, jack plane; C, block plane; D, trimming plane; E, wood chisels; and F, a utility knife. Courtesy Stanley Tools

A number of saws are also required for installing interior trim. While most saw cuts can be made with power saws, a small, fine-tooth **handsaw** will often be helpful. The longest of these saws is typically 26 inches long. However, the smaller sizes, ranging from 15 to 20 inches, are adequate for trim work. You want a fine-tooth saw. Typically, these saws have 9 or 12 points per inch. A small **miter box with a backsaw** is very useful for cutting small pieces of narrow molding. The saw is 14 inches long and has 13 points per inch. A large **miter box** with a saw built especially for it will handle almost any type of molding. A **dovetail saw** is a small, thin-blade, fine-tooth saw for accurate cutting that gives a very narrow kerf. A **wallboard saw** is about five inches long and has tough, coarse teeth. It is designed to cut gypsum drywall. Similar saws with fine teeth are available for cutting wood in tight places, and some even have a blade to cut nails and other forms of soft metal. The **coping saw** has a very thin blade and is used to cut curves in wood and plastic. In the following chapters, you will see it being used to cope molding. These saws are displayed in **2–5**.

2–5 Frequently used handsaws include: A, fine-tooth saw; B, small miter box with a backsaw; C, large miter box with a specially built saw; D, dovetail saw; E, wallboard saw; and F, coping saw. **Courtesy Stanley Tools**

FINISHING TOOLS

Files and **rasps** are used to remove small amounts of wood in order to make a difficult joint close properly (**2–6**). Files have very fine teeth and produce a fairly smooth surface. Rasps have large teeth and remove larger amounts of wood rapidly, but leave a rough surface. Both are available in flat, half-round, round, and triangular shapes, in a range of lengths.

A

B

C

D

E

2–6 Some of the most commonly used finishing tools include: A, files and rasps; B, file card; C and D, surform planes; and E, cabinet scraper. Courtesy Stanley Tools

Flat files are used to smooth convex surfaces (**2–7**). File down the curve. **Half-round files** are used to smooth concave surfaces (**2–8**). Slant the file a little and file down the curves.

Be certain to get good-quality strong file handles. The tangs of the file are sharp and can cause injury. A handle also gives better control of the file.

2–7 Convex surfaces are smoothed with a flat file that is moved down the curve.

2–8 Concave surfaces are smoothed with a half-round file that is moved down each half of the curve toward the bottom.

A **file cleaner** is needed to clean out the wood that sticks between a file's teeth (**2–9**). It has a fine wire brush on the back side of a wood handle.

2–9 Use a file cleaner to remove wood stuck between the file teeth. Clean the file frequently.

Surform tools have a series of perforated teeth much like a cheese grater (**2–6**). They remove wood quite rapidly. They can be used to smooth wood, aluminum, vinyl, fiberglass, and rubber. Recommendations for using surform tools are given in **2–10**.

Cabinet scrapers (**2–6**) are used to smooth the surface of wood. If it has slight irregularities or a raised grain, the scraper removes these with very thin shavings. **Paint scrapers** are designed to scrape off old paint after it has been softened with a paint remover. Be certain to test the paint for lead before trying to remove it. There are kits on the market you can use to test the paint for lead. Lead is found in paints in older homes. The lead particles are hazardous to your health if inhaled. Minimize the paint dust, wear a respirator, and remove all paint remains from the room.

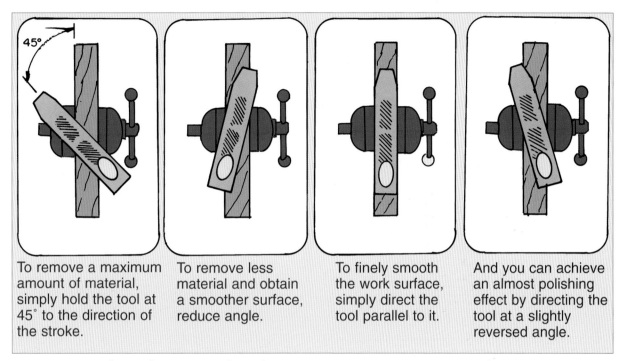

| To remove a maximum amount of material, simply hold the tool at 45° to the direction of the stroke. | To remove less material and obtain a smoother surface, reduce angle. | To finely smooth the work surface, simply direct the tool parallel to it. | And you can achieve an almost polishing effect by directing the tool at a slightly reversed angle. |

2–10 Recommendations for using a surform plane. **Courtesy Stanley Tools**

CLAMPS

There are a number of clamps available that can hold materials as they are being cut, smoothed, or installed. The **C-clamp (2–11)** is an inexpensive general-purpose clamp that is a must in every toolbox. A variation of this is the **spring clamp**. The jaws of this clamp are held closed by a heavy spring in the handle. Several types of **bar clamp** are available. These are used to hold large projects.

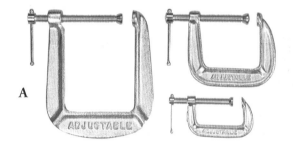

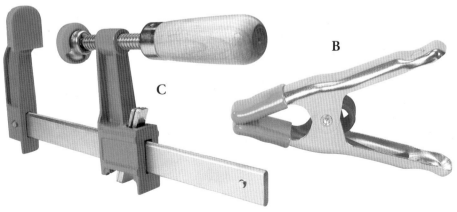

2–11 Frequently used clamps include: A, C-clamps; B, spring clamps; and C, bar clamps.
Courtesy Adjustable Clamp Company

MEASURING & LAYOUT TOOLS

Illus. **2–12** displays an assortment of measuring and layout tools. The two measuring tools that have been a staple of any toolbox for years are the **six-foot folding rule** and the **tape measure** (**2–12**). A 25-foot tape is most commonly used, but a 50-foot tape is sometimes helpful.

As trim layouts are made, it is sometimes helpful to locate a long level line along the wall to line up the trim. This line is made by snapping a chalk line at the desired level. If a plumb bob is placed on the end of the chalk line, it can be used to mark vertical locations.

Various squares can be used for checking right angles and layout work. The large **carpenter's square** has inch markings along the tongue and blade. The other markings are used to lay out roof rafters. Smaller sizes of the carpenter's square are available. The **combination square** has a head that moves along the blade. It is very useful when measuring a series of locations that all are the same length. The **bevel square** is adjusted to the desired angle and used to lay out or find the angle between two joining arms. The **try square** is a small tool used to lay out 90-degree angles and check for squareness. Metal **straightedges** are used to measure short distances and to draw straight lines.

2–12.

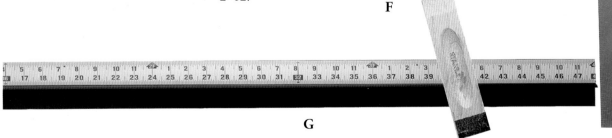

(continued on next page)

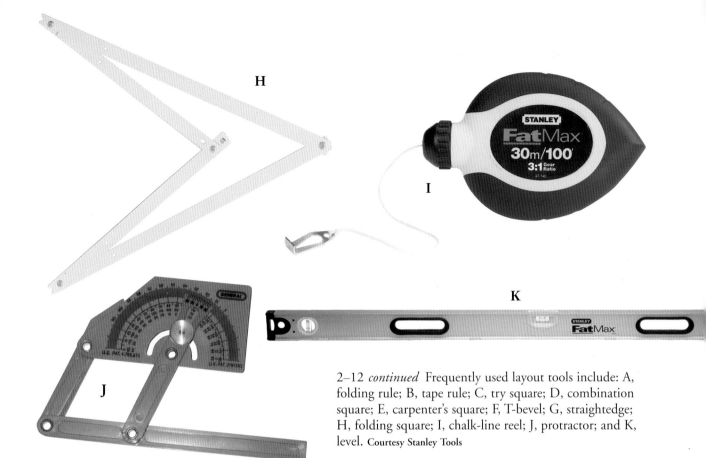

H

I

K

J

2–12 *continued* Frequently used layout tools include: A, folding rule; B, tape rule; C, try square; D, combination square; E, carpenter's square; F, T-bevel; G, straightedge; H, folding square; I, chalk-line reel; J, protractor; and K, level. **Courtesy Stanley Tools**

Careful marking of measurements is important. Very accurate markings can be made with a square or straightedge and a utility knife (**2–13**). An **electronic measuring tape** uses a steel tape and has a digital readout on the top. This means you no longer have to strain to read the markings on the tape (**2–14**).

A **digital protractor/angle finder** calculates bevel and miter angles for crown molding cuts. It measures the corner angle and computes the miter and bevel angles used to set the miter saw (**2–15**).

Levels are used to lay out horizontal and vertical surfaces. There are many types available. Some are as small as 9 inches long, while others 48 inches long are most useful when installing interior trim (**2–16**).

2–13 A utility knife used with a straightedge or square will make a very accurate mark locating the line of a cut. **Courtesy Stanley Tools**

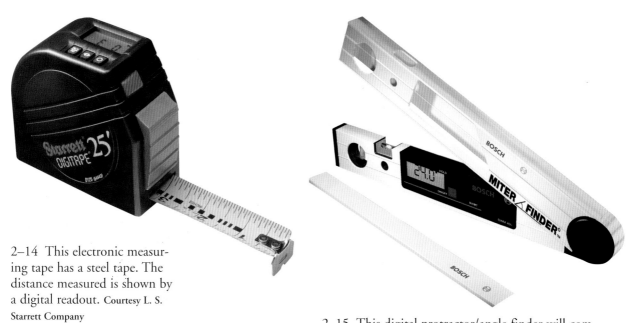

2–14 This electronic measuring tape has a steel tape. The distance measured is shown by a digital readout. Courtesy L. S. Starrett Company

2–15 This digital protractor/angle finder will compute the miter and bevel angles. Courtesy Robert Bosch Tool Corporation

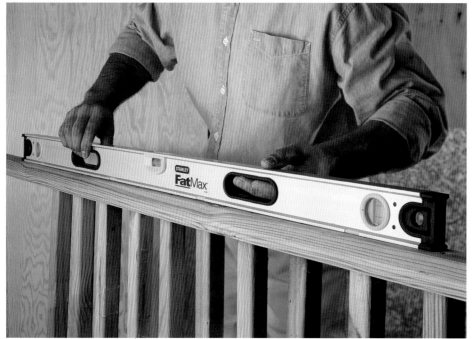

2–16 Long levels are required to check long horizontal and vertical members. Courtesy Stanley Tools

A **laser level** can be used to locate horizontal lines around a room (**2–17**). It is set on a tripod in the center of the room and, as it rotates, projects a line on the walls of the room. This is a good way to locate the line for a cornice molding at ceiling level (**2–18**). The laser level projects horizontal and vertical lines (**2–19**).

2–18 The laser level can line up chair rail, wainscoting, and vertical elements such as columns and pilasters. Courtesy DeWalt Industrial Tool Company

2–17 A laser level is mounted on a tripod and will locate horizontal and vertical lines. Courtesy DeWalt Industrial Tool Company

2–19 The laser level projects horizontal and vertical lines. Courtesy Stanley Tools

STRIKING TOOLS

Since many nail-type fasteners are power-driven, the main striking tool you will need is the curved **claw nail hammer** (**2–20**). Hammers with a rubber grip are less likely to slip when being used. They are available in several weights, with 16- and 20-ounce sizes commonly available.

Various types and sizes of **nail sets** (**2–20**) are very important. They are available in 4- and 5-inch lengths, with tips typically $\frac{1}{32}$, $\frac{1}{16}$, and $\frac{3}{32}$ inch in diameter. Smaller-diameter tips are used on smaller nail sets, and will make a smaller hole that needs to be filled when finishing the molding.

Often, it is necessary to remove old moldings, and a **pry bar** (**2–20**) is a great help. There are a number of designs and sizes available. They make it much easier to remove old molding and are less likely to damage it, which is important if it is to be reused.

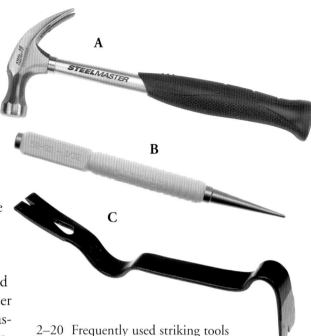

2–20 Frequently used striking tools include: A, curved claw hammer; B, nail set; and C, pry bar. **Courtesy Stanley Tools**

FASTENING TOOLS

The most frequently used fastening tool is the **screwdriver** (**2–21**). Standard screwdrivers are used for securing threaded fasteners with a single slot across the head. Phillips screwdrivers are used for screw heads that have an X-slot in them. Screwdrivers are available in a wide range of lengths and tip sizes. It is important to have a good set with various sizes.

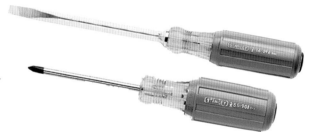

2–21 Standard and Phillips screwdrivers are an important part of the tool kit. **Courtesy Stanley Tools**

SAFETY EQUIPMENT

While installing interior trim is not as dangerous as framing carpentry and other construction trades, there are a number of safety-equipment items that are essential (**2–22**). Since there is considerable sawing that produces wood chips and sawdust, eye protection is mandatory. **Safety glasses** or **goggles** are typically

2–22

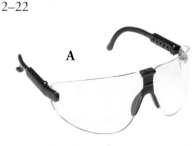

(continued on next page)

used. Sanding produces fine dust, so some form of **dust mask** is necessary. If heavy sanding is required or extensive amounts of paint or other finishing materials are used, a **respirator** is recommended. Some operations, such as shaping and planing, produce excessive noise, and noise-reduction **earmuffs** are required.

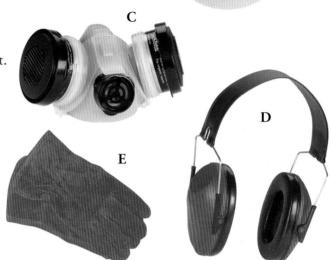

Gloves are useful when handling large amounts of materials. However, it is dangerous to wear them when operating power equipment.

When installing base and shoe moldings, it helps a great deal to wear **knee pads** (**2–23**). A number of designs are available.

2–23 Knee pads are a valuable item to use when installing baseboard and shoe moldings. Courtesy Fiskars Products

2–22 continued Required personal safety items include: A, safety glasses; B, dust mask; C, respirator; D, earmuffs; and E, gloves. Courtesy AOSafety

CLOTHING

Clothing should be comfortable and permit bending and other movement. Safety regulations require woodworkers to roll up clothing that could be caught in power tools, such as long sleeves. Scarves cannot be worn. Steel-toed shoes are recommended. However, athletic shoes such as those used for jogging are also available with steel toes, and they are more comfortable than heavy leather shoes.

A **tool belt** is helpful if the job requires you to carry a number of small tools (**2–24**). For many jobs, though, you can carry all the tools you need in your hand.

2–24 Tool belts keep the most frequently used tools handy. Courtesy Fiskars Products

Power Tools

There are excellent power tools available that make the job of installing interior trim faster and easier, and produce superior results. A finish carpenter or cabinetmaker will want to buy top-of-the-line professional tools. These have powerful motors and can perform heavy-duty work over the years. If you are a homeowner doing your own trimwork, a medium-quality, mid-priced power tool will be satisfactory. Avoid the low-cost, low-powered tools.

Power tools are available as stationary and portable units. **Stationary power tools** are mounted on a base that rests on the floor. **Portable power tools** are lightweight and are carried around as the work progresses. Portable power tools are operated by electricity, batteries, or compressed air. Select only tools with double-insulated electrical systems, to prevent electrical shocks. All tools should have guards on the cutting edges that provide good protection. Be certain the manufacturer has a reliable service and repair system.

GENERAL - POWER TOOL SAFETY RULES

1. Wear some form of approved eye and ear protection, and a dust mask when needed.
2. Use tools only with their guards in place.
3. Be certain the electrical cord and extension cord are not worn or damaged.
4. If the tool has a three-wire plug, be certain the source of electricity has a third wire ground and use a ground fault interrupter (GFI).
5. Remove all rings, bracelets, necklaces, etc., that may become caught. Long hair can also be a problem and should be confined with a hair net. Roll up your sleeves.
6. Avoid using a tool that is not in good condition.
7. Be certain the on-off switch operates properly.
8. Unplug the tool from the power source before changing a blade or other cutter.
9. Use only sharp saws and cutters. Dull tools can cause accidents.

10. After changing a saw blade or other cutter, recheck it before starting the tool to make certain it is properly seated and securely locked in place.
11. Work being cut must rest on a secure surface while being cut so it will not slip. If it is small, clamp it to the surface.
12. Avoid working when you are excessively fatigued.
13. Let the tool cut at its normal pace. Do not force it to speed up the cut. This can cause kickbacks.
14. After finishing a cut, let the guard close and the saw or cutter stop rotating before laying the tool down.
15. When leaving the machine unattended, lock the switch to prevent unauthorized use.

The Occupational Safety and Health Administration (OSHA) requirements for power tools are detailed in the publication *Hand and Power Tools, OSHA 3080*. Order it from OSHA Publication Office, 200 Constitution Avenue NW, Room N-3647, Washington, DC 20210.

TOOL PROTECTION

Power tools need their cutting edges protected. Some have guards that fold over the blade. Others, such as portable electric planes, need special protection of the cutting edges. Drills and bits need storage in special holders. Power drill bits, such as spade drills, have cutting edges that need protection. Power-tool manufacturers make plastic cases in which to store their portable power tools. They are designed for a particular tool, for example, a case for a router (**3–1**).

3–1 Store your portable power tools in the plastic cases provided by the manufacturer for that purpose. **Courtesy Hitachi Power Tools**

TABLE SAWS

The table saw is used for a variety of cuts, and produces more accurate results than the portable circular saw. The most common operations include ripping and crosscutting. When it has table extensions, it will accurately cut large panels such as plywood and particleboard. This tool can cause many accidents, so observing safety rules is very important (**3–2**).

3–2 This 10-inch table saw is a convenient size to use on the job. **Courtesy Delta International Machinery Corporation**

TABLE SAW SAFETY RULES

1. Read the instruction manual that comes with the table saw before operating it.
2. Do not adjust the saw while it is running.
3. Always wear safety goggles or safety glasses with side shields.
4. In some operations, a dust mask should be used.
5. The saw table and the floor around the saw should be kept free from sawdust and debris.
6. When cutting wide stock, such as plywood panels, use an auxiliary support to keep the panel level (**3–3**).

3–3 When cutting large panels, the table saw should have an extension table. Tables are available for both sides of the saw. **Courtesy Delta International Machinery Corporation**

7. Keep all guards in place. Repair them if they do not function properly (**3–4**).
8. The blade should project no more than ⅛ to ¼ inch above the work.
9. Do not reach over the blade nor place your hands within several inches of it.

3–4 When ripping stock on a table saw, always keep the saw guard in place and use the fence.
Courtesy Robert Bosch Tool Corporation

3–5 When crosscutting stock, always use a miter gauge. Never crosscut free-hand.
Courtesy Robert Bosch Tool Corporation

10. Never cut a piece of stock freehand. Always use a miter gauge for crosscutting and the rip fence for ripping (**3–4** and **3–5**).

11. Do not stand directly behind the blade. Kickbacks do occur and can cause serious injury.

12. When ripping narrow stock, use a push stick (**3–6**).

13. Be certain the rip fence is parallel with the blade. This reduces binding and burning and possible kickbacks.

14. Do not use cracked or burned blades.

15. Blades that wobble or vibrate must be discarded.

16. Be certain the blade is installed so that it rotates in the proper direction.

17. Place small, portable saws on a firm base at a convenient working height above the floor (**3–7**).

18. When cutting long boards, have someone help "tail off" the board as it leaves the saw. They should support the end of the board, but not pull on it.

19. Do not cut wet, cupped, or warped boards.

20. Do not use the rip fence as a stop when cutting short pieces to length. Use a stop block instead (**3–8**).

21. Do not leave a running saw unattended. If you have to leave it, shut it off.

22. Use the proper blade for the job at hand.

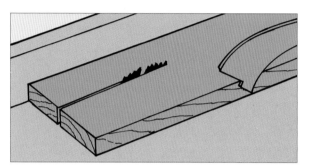

3–6 When ripping narrow stock, always use a push stick. The guard is not shown in this photo so the blade can be seen.

BLADE SELECTION

3–7 This saw stand holds a small 10-inch table saw. The legs fold under and it can be moved about on wheels. Courtesy Delta International Machinery Corporation

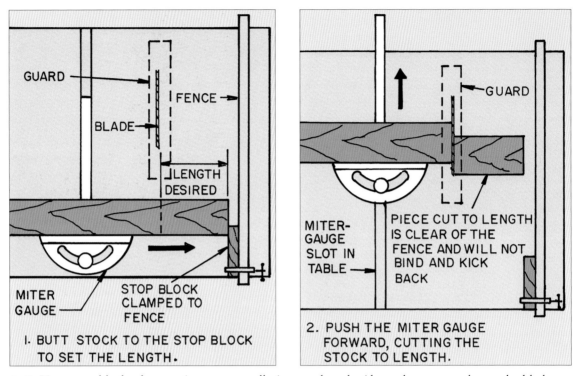

3–8 Use a stop block when cutting many small pieces to length. Always keep a guard over the blade when crosscutting.

There are many varieties of circular-saw blades available. The hole in the blade must match the diameter of the shaft on the saw. The following are the common types of blade:

A **crosscut blade** is used to cut stock across the grain. A **hollow-ground blade** is used for fine, smooth cuts. The teeth have no set, so it makes a very narrow kerf. A **rip blade** has large set, chisel-like teeth, and removes wood rapidly. It is used to cut with the grain. A **combination blade** has a combination of crosscut and rip teeth and is used for both ripping and crosscutting. A **plywood blade** is designed to cut plywood with a minimum of chipping, and leaves a smooth surface.

RADIAL ARM SAWS

A radial arm saw has a blade above the table that moves on an arm that extends over the table. A typical radial arm saw is shown in **3–9**. It can rip and crosscut as well as cut miters.

The blade is mounted directly onto the motor. The motor and blade are raised and lowered by a crank on the top of the column. They can also move horizontally by gliding along the arm. The blade must be installed with the teeth pointing down toward the table and rotating toward the fence when crosscutting stock (**3–10**). When ripping stock, the blade is set parallel with the fence. The stock is fed into the rotation of the blade (**3–11**). Flat and compound miters can also be cut with this tool.

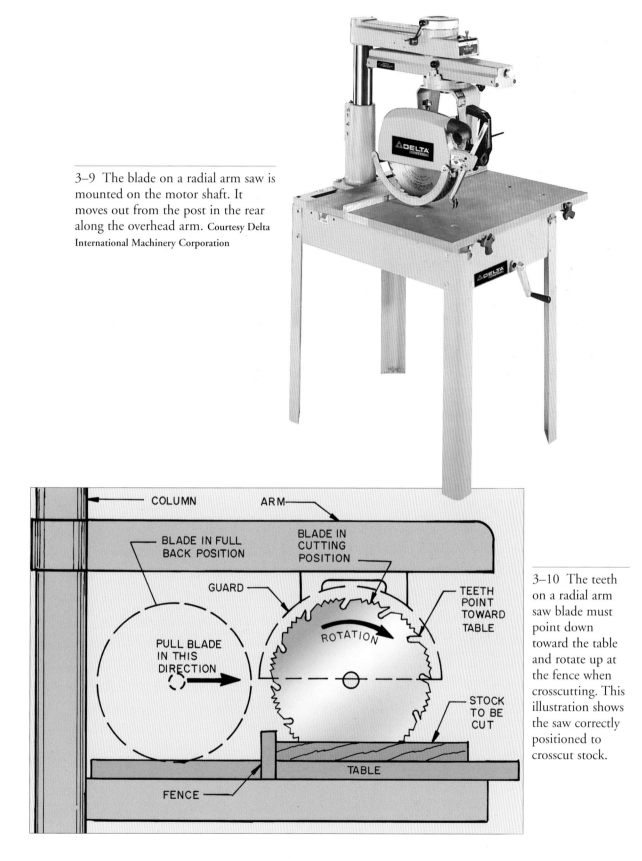

3–9 The blade on a radial arm saw is mounted on the motor shaft. It moves out from the post in the rear along the overhead arm. Courtesy Delta International Machinery Corporation

COLUMN

ARM

BLADE IN FULL BACK POSITION

BLADE IN CUTTING POSITION

GUARD

PULL BLADE IN THIS DIRECTION

ROTATION

TEETH POINT TOWARD TABLE

STOCK TO BE CUT

TABLE

FENCE

3–10 The teeth on a radial arm saw blade must point down toward the table and rotate up at the fence when crosscutting. This illustration shows the saw correctly positioned to crosscut stock.

RADIAL ARM SAW SAFETY RULES

1. Read the instruction manual that comes with the radial arm saw before operating it.
2. Never operate the saw unless all guards are in place.
3. Always wear safety goggles or safety glasses with side shields and, in some operations, a dust mask.
4. Be certain the blade is installed so that it rotates in the proper direction. If it is not, the saw will tend to run out toward the operator.
5. When ripping, be certain to feed the stock into the blade from the proper direction (**3–11**).
6. When crosscutting, hold the stock firmly until the blade is clear of the stock.
7. Keep both hands clear of the blade at all times. Never reach across the front of the saw.
8. Always return the blade/motor assembly to the fully retracted position after crosscutting a board.
9. Set the anti-kickback device so that it is slightly below the surface of the stock.
10. When ripping, be certain the blade is parallel with the fence. If not, binding will occur. This will produce a dangerous kickback.
11. Be certain to use the spreader device when ripping to keep the stock from binding behind the blade after it has been cut.
12. When ripping stock, use a push stick to feed it past the saw. Also, make certain the blade is rotating toward you.
13. Never cut anything freehand. Always keep the stock tight to the fence.
14. Do not cut wet, cupped, or warped wood.

MITER SAWS

The miter saw is used to cut miters and to crosscut narrow stock. It resembles the radial arm saw except that the blade/motor unit swings up and down on a pivot and may be adjusted to swing right and left at various angles (**3–12**).

Miter saws typically have 10- or 12-inch-diameter blades. Before selecting one, consider the sizes of molding it will cut when set at vari-

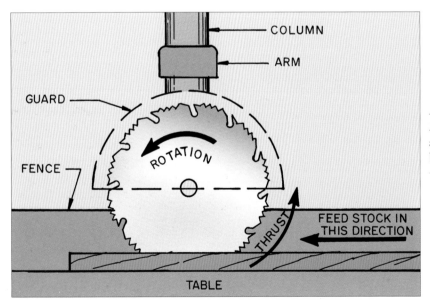

3–11 When ripping with a radial arm saw, feed the stock into the blade from the side where the teeth are rotating upward.

3–12 The miter saw is used to make square flat- and compound-miter cuts. **Courtesy Delta International Machinery Corporation**

ous angles. For example, a 10-inch blade will typically cut a 4¼-inch crown molding angled against the fence, while a 12-inch blade will cut crown molding that is 6⅝ inches wide (**3–13**).

The miter saw is probably the most frequently used power tool for installing interior trim. Since the various moldings must fit with almost perfect joints, the saw fence and blade relation-

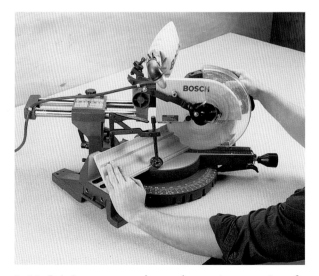

3–13 It is important to know the cutting capacity of a miter saw. **Courtesy Robert Bosch Tool Corporation**

ship must be checked occasionally to ensure accurate cuts. Do the following:

1. Check the blade to be certain it is perpendicular to the fence. Unplug the machine from the power source and lower the blade. Check the blade with a square as shown in **3–14**. Adjust it as recommended by the manufacturer.

3–14 A square is being used to check if the blade is perpendicular to the fence. Before doing this, unplug the machine.

2. Check to see that the blade is perpendicular to the table. Unplug the machine from the power source. Check the blade with a square as shown in **3–15**. Adjust it as recommended by the manufacturer.

3–15 The base of a combination square is being used to check if the blade is perpendicular to the table. Before doing this, unplug the machine.

3. Check the fence to see if it is perpendicular to the table (**3–16**). Adjust it as recommended by the manufacturer.

3–16 The fence must be perpendicular to the table. Check it with a square.

4. When cutting compound miters, check to be certain the blade is set on a 45-degree angle (**3–17**). Unplug the machine from the power

source and check the blade with the base of a combination square or some other fixed 45-degree-angle tool. If the blade angle is not correct, adjust it as recommended by the manufacturer.

Most miter saws will have adjustments to regulate the downward travel of the blade. The downward travel should allow the blade to cut through the stock but not hit the metal surface below the cutting slot. There are screws that control the downward movement. **Disconnect the machine from the power source before making this check.**

MITER SAW SAFETY RULES

1. Read the instruction manual that comes with the miter saw before operating it.
2. Always leave all guards in place.
3. Be certain the blade revolves down toward the table.
4. Since the saw swings down, it is necessary to be especially alert and keep your hands well clear of the cutting area.
5. Always hold the stock firmly against the fence (**3–18**). Never try to cut freehand.

3–17 Check the 45-degree blade setting to verify that the angle set by the stop is accurate.

3–18 Always hold the stock against the fence. Never cut stock that is resting loose on the table.
Courtesy Robert Bosch Tool Corporation

6. As soon as a cut is complete, release the electric switch.
7. Be certain the spring that lifts the saw to an upright position is properly adjusted. Do not use the saw if it will not automatically rise to the up position when released.
8. Secure the miter saw to a firm base that holds it at a comfortable working height and provides support for long moldings (**3–19**).

3–20 This electric handsaw has a variable-speed control and blades with fine teeth that produce a smooth-finish cut. **Courtesy Robert Bosch Tool Corporation**

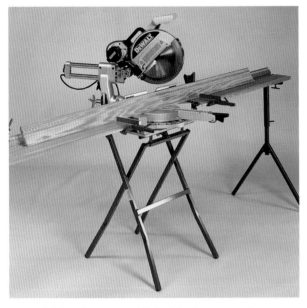

3–19 This miter-saw table holds the saw at the proper working height and supports long pieces of molding. **Courtesy Colonial Saw Company**

3–21 When used with a special miter jig, the precision electric handsaw can cut at various angles. **Courtesy Robert Bosch Tool Corporation**

PRECISION POWER HANDSAWS

A small, lightweight precision handsaw makes it possible to quickly and accurately cut handrails, door and window casings, baseboards, crown moldings, and other such items. The saw has several blades with ultra-fine teeth, providing a smooth cut edge (**3–20**). The saw can be used in connection with a special miter jig to cut moldings at 90 degrees, as well as for various other mitering applications (**3–21**). This makes it especially useful for finish and trim carpentry.

**PRECISION POWER HANDSAW
SAFETY RULES**

1. Review the manufacturer's operating instructions.
2. Clamp or hold the stock securely to the fence.
3. Keep your hands clear of the cutting area.
4. Move the saw down, slowly letting it cut at its own speed. Do not force it to cut faster.
5. As soon as the cut is finished, release the switch and then raise it to the up position.

JOINTERS

Jointers are used to straighten the edges and flatten the faces of cupped boards. They are also used to cut chamfers, bevels, and rabbets. The surfaces produced are smooth and flat and require less sanding than saw-cut surfaces (**3–22**).

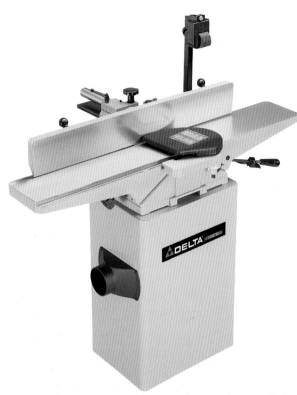

3–22 A jointer is used to straighten and smooth the edges and faces of stock. **Courtesy Delta International Machinery Corporation**

JOINTER SAFETY RULES

1. Read the instruction manual that comes with the jointer before operating it.
2. Keep your fingers and hands at least four inches away from the cutterhead.
3. A cutterhead guard should be used at all times.
4. The maximum depth of cut in one pass is 1/16 inch.
5. Stock must be at least ten inches long before it can be face- or edge-jointed.
6. Stock that is less than one inch wide should not be edge-jointed.
7. Stock that is less than 1/2 inch thick should not be face-jointed.
8. Use a push shoe or push stick whenever possible.
9. Stock must be held flat on the table and firmly against the fence.
10. The outfeed table should never be adjusted—unless you are a skilled operator.
11. Disconnect the power before getting your hands close to the cutterhead.
12. Do not cut boards that have knots.
13. The depth of cut is regulated by lowering the infeed table (**3–23**). Do not try to take too deep a cut. Several lighter cuts will produce the best results. Usually a cut depth of 1/16 inch is used.
14. To make an edge square with the face of the board, set the fence at 90 degrees to the table. To cut bevels and chamfers, slant the fence (**3–24**). Pass the board over the cutter one or more times as needed to get the depth of cut you want.

3–23 The depth of cut on a jointer is set by lowering the infeed table.

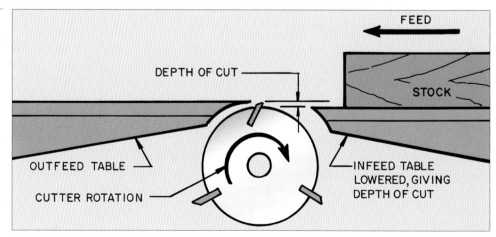

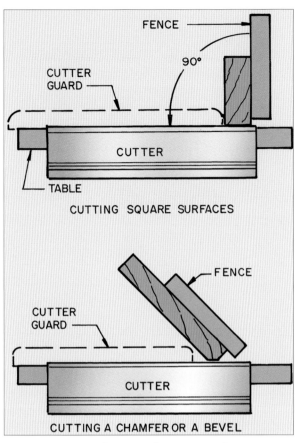

FENCE

90°

CUTTER GUARD

CUTTER

TABLE

CUTTING SQUARE SURFACES

FENCE

CUTTER GUARD

CUTTER

CUTTING A CHAMFER OR A BEVEL

3–24 The jointer fence can be set at 90 degrees to the table for square cuts, or angled to form a chamfer or bevel.

3–25 This heavy-duty saber saw will make straight and curved cuts in thick stock. Courtesy DeWalt Industrial Tool Company

3–26 A saber saw is widely used to cut curves in plywood and other paneling products. Courtesy DeWalt Industrial Tool Company

SABER SAWS

The saber saw (also known as a bayonet saw) is a portable electric saw with a reciprocating blade that is primarily used to cut curves and make internal cuts. It can also be used for cross-cutting, ripping, and mitering, but it is not as accurate as other types of saw. A typical saber saw is shown in **3–25**. Light-duty saber saws are used to cut thin materials such as plywood sheets (**3–26**). Heavy-duty saber saws can be used to cut 2-inch stock. Several brands will cut stock up to 5 inches thick.

Manufacturers provide a variety of blades for various purposes. Thin wood is best cut with a fine-tooth blade such as one with 12 teeth per inch. A blade with 10 teeth per inch is best on thicker panels. If cutting 2-inch stock, a blade with 6 teeth per inch will provide the fastest cut. Some blades will cut metal and plastics and can even be used to cut through nails. Thin, hard metals and very thin woods are best cut with a blade that has 24 teeth per inch. Thicker, softer metals require a blade with 14 teeth per inch. A blade should always have two teeth in contact with the material.

Internal cuts can be started by drilling holes at the corners of the area to be removed. The blade is then lowered into the hole and the cut proceeds in a normal manner.

SABER SAW SAFETY RULES

1. Read the instruction manual that comes with the saw before operating it.
2. Unplug the saw before changing blades.
3. Before cutting, check to be certain that the blade is securely held in the chuck.
4. As you cut, keep the cord out of the way of the blade.
5. When beginning a cut, place the base firmly on the wood's surface before starting the blade into the wood.
6. If the blade gets stuck, turn off the power and slide it clear.
7. When plunge-cutting, use blades specifically designed for that purpose.
8. The blade is very hot after cutting, so do not touch it.
9. Do not force the saw to cut faster than normal.
10. Do not use dull blades.
11. Be sure you are holding the wood so that it will not slip. Clamp it if necessary.
12. Let the saw get to full speed before starting to cut.
13. Keep all fingers on top of the board.

PORTABLE CIRCULAR SAWS

The portable circular saw is probably the most dangerous of the portable power tools. Carelessness and improper use can cause serious accidents.

A typical saw is shown in **3–27**. Popular saw sizes use 7¼- and 8¼-inch-diameter blades.

3–27 The portable circular saw is available in a range of blade sizes and horsepower. **Courtesy Robert Bosch Tool Corporation**

However, the saws are available in other sizes. They are used for crosscutting and ripping trim, solid lumber, plywood, and other paneling products. There are many types available, and the purchaser needs to consider possible uses when making his/her selection. Factors to consider are the blade size, whether the saw has the proper power for the blade, the quality of the guards, whether the saw has provision for grounding and a brake on the blade, the saw's weight, and type of drive (helical or worm drive).

The circular saw cuts from the bottom of the material, producing a smoother cut at the bottom. This means the best surface of the stock should be placed down, and the cut made from the other side.

The angle the blade makes with the surface can be adjusted from 90 to 45 degrees. This permits bevel cutting on ends and edges. To do this, the base is unlocked and pivoted to the angle desired (**3–28**).

When installing a blade, be certain it is placed on the arbor so that the teeth cut up from the bottom (**3–29**). Usually, the blade is marked with an arrow on the outside face that indicates the direction of rotation.

3–28 The portable circular saw can make cuts from 90 to 45 degrees. Courtesy DeWalt Industrial Tool Company

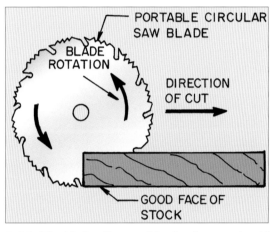

3–29 The blade of a portable circular saw should be installed so the blade cuts up from the bottom of the stock.

Crosscutting can be done freehand, if accuracy is not important. A straightedge clamped to the stock will provide a fence against which the saw can slide. This provides a more accurate cut (**3–30**).

Ripping can also be done freehand by keeping the blade cutting along a line drawn on the surface of the wood. Long straightedges or straight stock can be clamped to the surface to produce a more accurate cut.

PORTABLE CIRCULAR SAW SAFETY RULES

1. Read the instruction manual that comes with the portable circular saw before operating it.
2. Do not use blades that are dull, cracked, or wobbly.
3. Keep the extension cord clear of the work. Be certain the cord is long enough to let you finish the cut. A short cord may jerk the saw back, causing a kickback.
4. The piece being cut must be clamped or otherwise held firmly in place.
5. Never hold a piece of wood in your hand while trying to cut it. It must be held firmly against some stationary object.
6. Do not try to cut small pieces of wood.
7. Set the depth of cut so that the blade protrudes no more than ⅛ inch below the wood being cut.
8. Never place your hands below the wood being cut.
9. Allow the blade to reach full speed before starting the cut.
10. A bind between the saw and wood could cause a kickback. If the saw binds, release the switch immediately. Put a wood wedge in the kerf to hold it open before trying to continue the cut.

11. Do not use the saw if the guards are not working.
12. After finishing a cut, let the guard close and let the blade stop before moving it to another position.
13. Never use a dull blade. This will cause kickbacks and burning of the blade.
14. Do not force the saw to cut faster than its normal pace.
15. Remove pitch and resin buildup from the blade.
16. Do not cut wet wood. This increases friction and loads the blade with wet sawdust.
17. When cutting large pieces, have a means of preventing the pieces from falling when the cut is completed. A helper is often a good solution.

PORTABLE ELECTRIC PLANERS

Portable electric planers are used to smooth the edges of stock and the faces of narrow boards. They will cut chamfers, rabbets, butt joints, and tenons. The width of the cutter is generally about three inches (**3–31**), and a typical depth of cut in a single pass is .059 inch.

To use the electric planer, set the desired depth of cut in the manner specified by the manufacturer. Do not set it too deep. It is better to take two light cuts than one overly deep cut. The fence slides along the face of the board and keeps the planer moving in a straight line. Start the motor, and let it reach full speed. Place the front shoe on the board, and slide the planer along the surface. Hold the planer as specified by the manufacturer. At the start of the cut, keep most of the downward pressure on the front shoe. As you reach the end of the surface, keep most of the pressure on the rear shoe (**3–32**). Plane in the direction of the grain.

In addition to planing edges square and smoothing flat surfaces (**3–33**), the electric planer will plane bevels and chamfers when used with a bevel fence guide (**3–34**).

PORTABLE ELECTRIC PLANER SAFETY RULES

1. Read the instruction manual that comes with the plane before operating it.
2. Be certain the cutters are sharp and installed as recommended by the manufacturer.
3. Allow the cutterhead to reach full speed before starting a cut.
4. Use shallower cuts on harder woods.
5. Hold the tool as recommended by the manufacturer. Large planers require two hands. Small block planers can be held with one hand.

3–31 This portable planer will take a cut .059 inch deep and 3¼ inches wide.
Courtesy Robert Bosch Tool Company

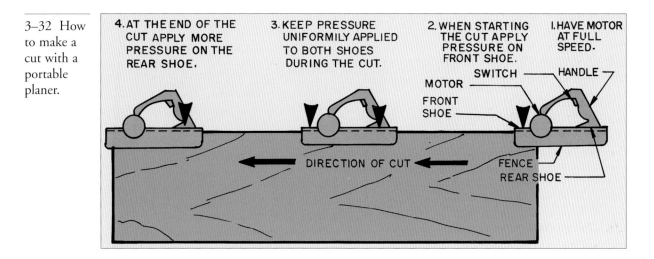

3–32 How to make a cut with a portable planer.

4. AT THE END OF THE CUT APPLY MORE PRESSURE ON THE REAR SHOE.

3. KEEP PRESSURE UNIFORMILY APPLIED TO BOTH SHOES DURING THE CUT.

2. WHEN STARTING THE CUT APPLY PRESSURE ON FRONT SHOE.

1. HAVE MOTOR AT FULL SPEED.

SWITCH HANDLE

MOTOR

FRONT SHOE

DIRECTION OF CUT

FENCE

REAR SHOE

3–33 The portable electric planer can smooth flat surfaces and make them square with the edge of the stock. **Courtesy Robert Bosch Tool Corporation**

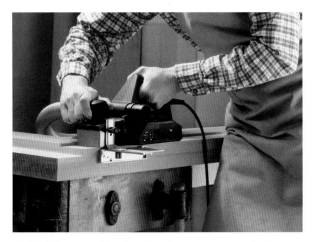

3–34 The edge of this door is being beveled with an electric planer. The beveled edge keeps it from striking the doorjamb as the door is closed. **Courtesy Robert Bosch Tool Corporation**

6. Be certain the stock is held securely. If not, the plane will throw it back.
7. Keep both hands on top of the planer.
8. Do not feed the plane faster than it seems to cut easily.
9. When possible, avoid cutting knots, especially on deep cuts.

PORTABLE ELECTRIC DRILLS

Portable electric drills are available in a wide range of sizes and features. Some have 120-volt electric motors and can be plugged into outlets; others, called cordless drills, are battery-operated. Typical electric drills are shown in **3–35** and

3–35 This is a typical small, convenient-size portable electric drill. It operates on 120-volt current. **Courtesy DeWalt Industrial Tool Company**

3–36. Most small drills have a pistol-grip handle and can be operated with one hand. Larger, more powerful drills have a pistol grip and a second handle to control the torque they produce (**3–37**).

Electric drills operating on 120-volt current are classified by chuck size and by the amperage of their electric motors. The chuck size is the maximum-diameter drill shank it will hold. The most frequently used sizes are ⅜ and ½ inch. Typical amperage ratings run from 3 to 10 amps. Battery-operated cordless drills are specified by chuck size and the available voltage. Common chuck sizes are ⅜ and ½ inch, and voltages of 7 to 13 are typical. The chuck may require a key to tighten it on the drill or be a keyless type.

When using straight-shank twist drills, insert the round shank into the chuck, but do not let any of the twisted flutes enter. Drills with no flutes, such as a spade bit, are inserted into the chuck as far as they will go (**3–38**). It is recommended that the chuck be tightened by inserting the key and tightening it at all three holes in the chuck. The bit can be released by inserting the key into any one of the holes.

When drilling holes, be certain to keep the drill perpendicular to the work. If it drifts off at an angle, the bit may break or bind in the work.

3–37 This powerful electric drill operates off 120-volt current and will handle heavy jobs. Notice that it has two handles for extra control. Courtesy Hitachi Power Tools

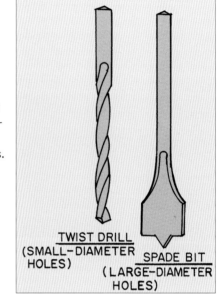

3–38 Twist drill bits and spade bits are commonly used in portable electric drills. The spade bits produce large-diameter holes.

TWIST DRILL (SMALL-DIAMETER HOLES)

SPADE BIT (LARGE-DIAMETER HOLES)

3–36 This portable electric drill operates off a battery, and provides the convenience of being free from an electrical cord.
Courtesy Robert Bosch Tool Company

When a large bit in a powerful drill binds, the tool can twist out of your hands, causing injury.

Some drills operate at various speeds, and can reverse their direction of rotation. They can be equipped with a screwdriver bit and used to drive screws. This is done at a very low speed of rotation.

PORTABLE ELECTRIC DRILL SAFETY RULES

1. Read the instruction manual that comes with the electric drill before operating it.
2. Do not use the drill if the power switch does not operate properly.

3. Remove the chuck key before starting the drill.
4. Always hold the tool securely.
5. Do not force the drill bit into the work.
6. Use sharp drill bits.
7. If the drill bit binds in the work, immediately release the switch. Free the bit from the work before proceeding.
8. When the drill is about to break through the back of the stock, reduce pressure.
9. Drill bits get very hot, so do not touch them immediately after finishing a hole.
10. Keep drill bits clean and free of wood chips or resin deposits.
11. Be certain the work being drilled is firmly supported.

PORTABLE ROUTERS

The portable router is used to make decorative cuts and joinery. It consists of two major parts: the motor and the base (**3–39**). On the end of the motor shaft is a collet that holds the router bits. The height of the motor can be adjusted inside the base. This adjustment sets the depth of cut. A lock binds the motor and base together when the proper depth has been set. For general uses, a one-horsepower router is adequate.

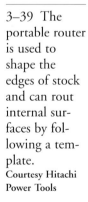

3–39 The portable router is used to shape the edges of stock and can rout internal surfaces by following a template.
Courtesy Hitachi Power Tools

A finish carpenter will most likely want a 1½ to 2½ horsepower router so that deeper cuts may be made.

An excellent way to shape custom molding is to mount a router below a router table. The bit extends above the table and the stock is guided by a fence. There is considerable information in Chapter 1 concerning routers in general, and this technique is discussed there.

ROUTER SAFETY RULES

1. Read the instruction manual that comes with the router before operating it.
2. Deep cuts can overload the router. Limit cuts to ⅛ to ³⁄₁₆ inch.
3. Before changing the bit, unplug the power plug.
4. Stock to be routed should be clamped to sawhorses or a workbench.
5. Hold portable routers with two hands using the handles provided on the machine.
6. Never place your hands near the rotating bit.
7. When routing an outside edge of the stock, move the router in a **counterclockwise** direction.
8. Be certain the bit is installed correctly.
9. Do not let the rotating bit become tangled in your clothing, and wear long hair in a hair net.
10. Wear eye protection and a face shield.
11. Do not use dull router bits.

PLATE JOINERS

Plate or biscuit joiners (**3–40**) use a small cutter to cut short, circular kerfs in the edges of the wood to be joined. Pressed-wood splines, called plates or biscuits (**3–41**), are glued into these kerfs, and hold the wood parts together. They are commonly used on edge-to-edge, butt, and miter joints (**3–42**). On interior trimwork, plates are an excellent way to secure miters on the corners of door and window casings so they will not open up over time.

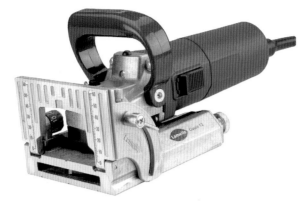

3–40 Plate or biscuit joiners cut concave slots in wood members that are to be joined with biscuits or plates. **Courtesy Colonial Saw Company, Inc.**

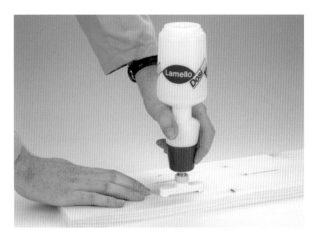

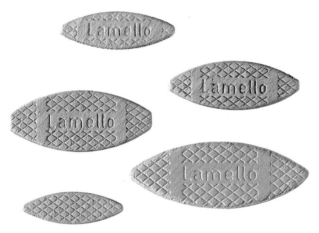

3–41 These pressed-wood plates are glued into the grooves cut in the wood. **Courtesy Colonial Saw Company, Inc.**

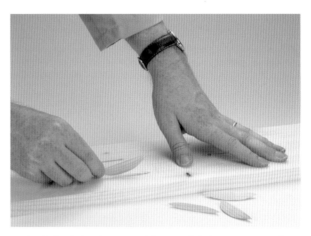

3–42 The joint is created by machining a groove for the plate in each of the pieces to be joined, putting glue in the grooves, inserting the plates, pulling the pieces together, and clamping them until the glue dries. **Courtesy Colonial Saw Company, Inc.**

PLATE JOINER SAFETY RULES

1. Read the instruction manual before using the plate joiner.
2. Be certain the saw blade is installed as directed by the maufacturer.
3. Clamp the work to be slotted to a steady workbench.
4. Keep the retractable blade guard over the blade.
5. Be certain the depth stop is set to the desired depth and locked tightly.
6. Hold the plate joiner by the handles provided. Do not get your fingers below the wood edge being slotted.
7. If the motor stalls immediately, turn off the switch.
8. When finished cutting, turn off the switch. Remove the blade from the slot when it stops turning.

LAMELLO WOOD REPAIR SYSTEM

Frequently a piece of molding will have a damaged spot. Small holes may be repaired by filling them with a wood filler. However, larger repairs warrant a wood patch. This is a more permanent type of repair.

The damaged area can be cut out using a patch-cutting machine (**3–43**). This tool can make openings for various size patches, which are made with a patch-making machine (**3–44**). A machine for cutting openings for round patches is also available. Glue the patch into the recessed groove. After the glue has dried, plane the patch flush (**3–45**).

BELT SANDERS

The two portable sanders commonly used for finishing are belt- and pad-type sanders. The belt sander is primarily used for removing large amounts of wood from flat surfaces (**3–46**). These are available in a variety of sizes. Their size is determined by the length and width of the belt, 3 × 21 inches and 4 × 24 inches being the most widely used. Some have a dust bag attached to reduce the amount of dust in the air.

USING A BELT SANDER

Before starting the belt sander, be certain the belt is properly installed. The sander has tracking knobs on each side that need frequent adjustment to keep the belt running straight on the pulleys.

Clamp the stock to be sanded, or it will be thrown across the room when the moving belt is placed against it. Place the sander flat on the wood, grip the handles firmly, and turn on the power. Move the sander immediately, following the pattern in **3–47**. The strokes should overlap and be short. Always sand with the grain.

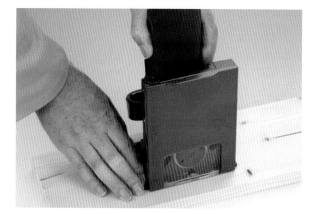

3–43 This patch-cutting machine cuts away a small damaged area, which is then filled with a wood patch. Courtesy Colonial Saw Company, Inc.

3–44 Patches made to repair damaged areas are prepared by the Lamello wood-grooving machine. Courtesy Colonial Saw Company, Inc.

3–45 After the patch has dried, plane it flush with the surface of the board. Courtesy Colonial Saw Company, Inc.

3–46 This belt sander uses a continuous belt. When the belt has a coarse abrasive, it can remove large amounts of material. **Courtesy Hitachi Power Tools**

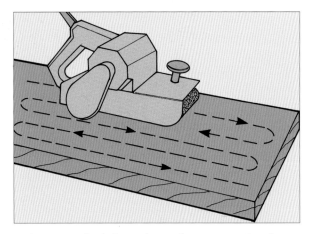

3–47 Move the belt sander in this pattern. Overlap the strokes.

SANDER SAFETY RULES

1. Read the instruction manual that comes with the belt sander before operating it.
2. Unplug the power cord before changing the abrasive belt or pad.
3. Be certain the switch is in the OFF position before plugging in the power cord.
4. Clamp pieces to be sanded to a workbench or sawhorses.
5. Keep both hands on the handles of the sander (**3–48**).
6. Keep the power cord out of the way. Some people run it over their shoulders while working.
7. Do not set the sander down while it is still turning.

3–48 Hold the belt sander firmly with both hands. **Courtesy Robert Bosch Tool Company**

FINISHING SANDERS

Finishing sanders are used to produce the final finished surface that is ready for painting or staining. They are used after a belt sander has removed any defects. Moldings and paneling seldom require the use of a belt sander. If they are in that bad a condition, they should be discarded.

The vibrating finishing sander has a square pad (**3–49**). Some models can be switched to run with an orbital, vibrating, or straight-line action (**3–50**). The orbital action cuts in a circular motion and is used for various types of finish sanding. This is followed up by using the straight-line action to produce the final smooth finish. The vibrating action cuts across the grain.

The sanding pad is covered with abrasive paper that is held to the pad with clamps at each end. Since this is a finish sander, fine-grit abra-

3–49 A vibrating finishing sander uses a fine abrasive paper and produces a very smooth finished surface.
Courtesy Hitachi Power Tools

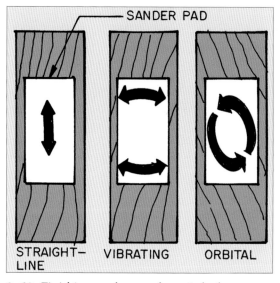

3–50 Finishing sanders can be switched to move in vibrating, orbital, or straight-line patterns. Always finish with the straight-line pattern.

sive papers are generally used. Do not bear down on the sander to try to speed up the removal of a defect: Use a coarser paper, or go back to the belt sander instead.

Refer to Sander Safety Rules above before using a finishing sander.

POWER NAILERS & STAPLERS

Power nailers speed up the work of the finish carpenter (**3–51** and **3–52**). They have a variety of uses, such as installing doors, casing, baseboard, and cabinetwork. Power nailers are driven by an air compressor. A cordless type is available. Some nailers drive small brads, as is required when installing small trim and moldings. Brad sizes range from ⅝ to 1⅝ inches. A finishing nailer is used for installing light and heavy trim, paneling, and stair parts. Finishing nails range from 1 to 2 inches in length. Heavy-duty nailers are used by framing carpenters and drive nails up to 16d.

Power staplers are also used for interior work, but are only used in places where the staples will not be seen.

3–51 Power nailers are used to install crown moldings, casings, baseboards, chair rails, and other interior finish woodwork.
Courtesy Senco Fastening Systems

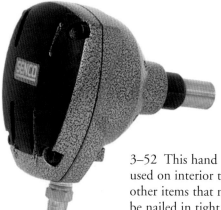

3–52 This hand nailer is used on interior trim and other items that need to be nailed in tight places.
Courtesy Senco Fastening Systems

POWER NAILER & STAPLER SAFETY RULES

The following are safety rules to observe when using a power nailer and stapler:

1. Before operating the tool, read the owner's manual and observe all operating recommendations.
2. Treat the tool as you would a gun. *Never* point it at anyone, even as a joke. The projected fastener can cause serious physical damage.
3. Keep both hands behind the nail-ejecting tube.
4. Keep your feet and legs clear of the tool.
5. Wear safety glasses.
6. Use only the fasteners designed for the tool and recommended by the manufacturer.
7. Keep the tool tight against the surface being fastened. Do not let it bounce.
8. Maintain the recommended air pressure.
9. When making repairs or adjustments, disconnect the tool from the air line.
10. Do not use a nailer that discharges a nail when the end of the tool is not against a surface.

AIR COMPRESSORS

In order to run an air-driven tool, it is important to have an air compressor that supplies a consistent airstream at the required pressure (80 to 100 psi). The air supply must also provide the required lubrication to the power tool as specified by the manufacturer. Water condenses in the air tank, and must be drained regularly . Filters are used to keep the air clean and remove abrasive sludge created by rust and dust in the air-supply system. The air-intake filter must be cleaned frequently (**3–53**). The air compressor must have the capacity to operate the tools that will be connected to it.

3–53 Since many power tools used for installing interior trim operate on compressed air, an air compressor is a vital part of the equipment needed on the jobsite.
Courtesy Thomas Industries.

CHAPTER 4

Window Casing

The type of trim installed around the windows is a major factor in defining a room. The size and style of casing directly influence the reaction of those who enter it. Casing should also reflect the architectural style of the house (4–1).

There are many styles of casing available. Some are described in Chapter 1. When it arrives on the job, the casing should be laid flat in a dry room and covered to keep it clean. Generally, the same style casing is used on windows and doors.

STYLES OF INTERIOR WINDOW CASING

Possibly the most commonly used style is some form of a mitered frame. The window can be framed with casing on all four sides, or only on three sides with a stool[1] and apron[2] at the bottom (4–2).

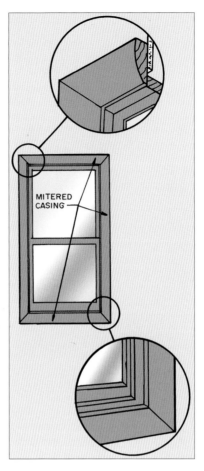

4–1 The choice of window casing, including the profile and species of wood, influences the appearance of the entire room.
Courtesy Weather Shield Manufacturing Company

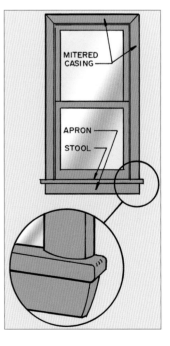

4–2 Mitered window casing can be installed with or without a stool and apron.

[1] A stool is the bottom interior trim of a window frame that forms a flat, narrow shelf. Although actually a different piece, the stool is often called the sill. The stool is the sill cap.
[2] An apron is the lower member under the sill of the interior casing of a window.

Another style is a rectangular casing that joins the head casing in a butt joint. A variation of this is made by adding various moldings. A few examples are shown in **4–3**. One type has a flat head casing extending beyond the side casing. Another type uses some form of molding to enclose the trim. Rosettes (also referred to as corner blocks) are typically used when the designer is reproducing a traditional period style of house. Rosettes are available from various manufacturers.

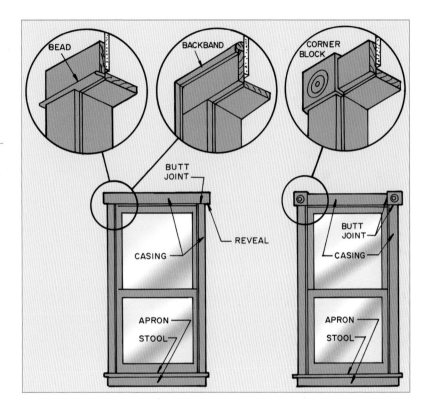

4–3 Butted window casing can have its head casing decorated with some type of molding. Rosettes (also referred to as corner blocks) also add to the appearance.

PREPARING TO INSTALL THE CASING

Before starting to cut and fit trim, check the window frame to see that it is plumb, level, and square. This is the time to make corrections. Check to see if the windows open and close smoothly. Sometimes the width of the exterior wall is greater than that of the window jamb. This makes it necessary to install a jamb extension. Jamb extensions are wood strips added to the frame to bring it flush with the drywall (**4–4**). Window manufacturers supply jamb extensions. The most commonly available types are a rectangular square-edged strip, or one with a tongue that glues into a groove in the window frame. Both types of extensions are glued and nailed in place.

Jamb extensions can also be made on the job. White pine is the recommended wood, because most windows are made from that material. If the extension used is wider than the wall, the width can be marked by drawing a line where it touches the drywall (**4–5**). The jamb extension is usually glued to the window frame. Narrow extensions can be held on by nailing or screwing them into the frame. Wider extensions may require blocking between them and the trimmer

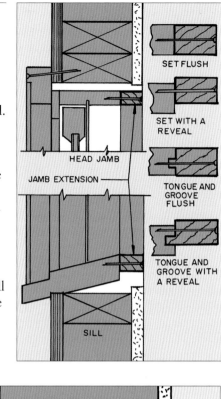

4-4 The square jamb extension can be set flush with the jamb or with a reveal. The reveal hides the crack. Some windows are manufactured with a groove to receive an extension with a tongue. Drill holes for the nails.

SET FLUSH

SET WITH A REVEAL

HEAD JAMB

JAMB EXTENSION

TONGUE AND GROOVE FLUSH

TONGUE AND GROOVE WITH A REVEAL

SILL

stud, and they are nailed to the stud. Careful work is required to get a good joint.

Before the casing is installed, insulation should be inserted into the space between the frame and the studs and header. Loose insulation can be inserted, but it must be kept loose and fluffy. Expanding foam insulation (which looks like shaving cream) is available in small pressurized cans. It is good for filling small cracks. Since it expands greatly, only a very small amount is required.

When vinyl windows are installed, the interior sides of the rough opening are finished with a quality wood-trim board to which the casing is nailed (4–6).

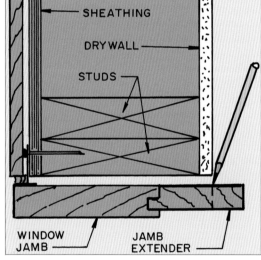

SHEATHING

DRYWALL

STUDS

WINDOW JAMB

JAMB EXTENDER

4–5 Mark the jamb extender and cut it so it is flush with the finished interior wall.

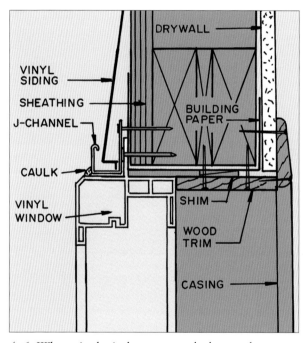

DRYWALL

VINYL SIDING

SHEATHING

J-CHANNEL

BUILDING PAPER

CAULK

VINYL WINDOW

SHIM

WOOD TRIM

CASING

4–6 When vinyl windows are used, the rough opening is finished with wood trim that runs flush with the drywall.

CUTTING MITERS

A miter joint is commonly used to form the corners of door and window casings. A miter joint is made when two pieces of material meet at a right angle; each is cut on a 45-degree angle (4–7). After a casing has the points marked locating the short side of the piece, the cut is made with a miter saw (4–8). Typically, this is a power-operated saw that will make accurate cuts.

It is wise to cut a miter on some scrap stock to check the saw settings. It should cut on a perfect 45-degree angle; however, the saw may need some adjusting to achieve this. Check the angles with a combination square (4–9). However, the real test of the miter is when you begin to install it. Changes may be necessary, as explained in the following section.

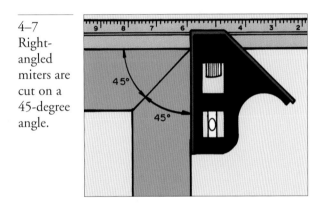

4–7 Right-angled miters are cut on a 45-degree angle.

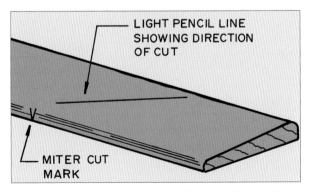

LIGHT PENCIL LINE SHOWING DIRECTION OF CUT

MITER CUT MARK

4–8 The miter is cut on the mark that is located by the reveal marks. A light pencil line indicates the direction of the cut.

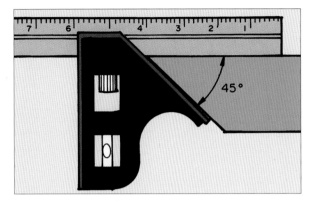

4–9 A miter can be checked with a combination square.

It is important that you check to see that the saw blade is running perpendicular to the fence, and the blade is set perpendicular to the table. (See Chapter 3 for more information on checking the miter-saw settings. Also, review the owner's manual. **Always disconnect the saw from the power before making the check.**) However, don't worry if the saw cuts the edge with a slight slope to the **back** of the casing, as this helps get the joint to close tightly.

When cutting miters using marks on the short side, place the marks next to the fence (4–10). Be certain to hold the casing firmly against the fence and table. Start the motor, and after the blade reaches full speed, lower it into the molding. Always keep the guard over the blade and wear eye protection.

SETTING THE MITER SAW & MAKING THE CUT

A scale for setting the angle of a cut is shown in **4–11**. Some saws only give the degree reading from 0 to 45 left and right. The point marked 0 is really 90 degrees with the fence, so when the blade is set on 0, it will cut the molding square on the end. To cut a 45-degree angle, move the pointer on the handle to 45 degrees. If some other angle is needed, say 60 degrees, move the pointer to the 30-degree mark. Since the 0 mark

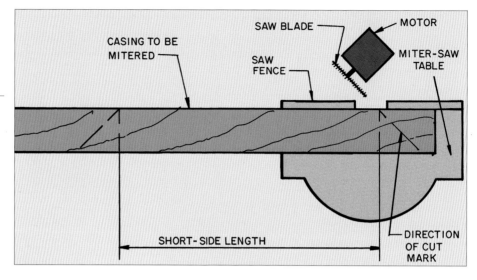

4–10 When cutting miters from the short side, place this side next to the miter-saw fence.

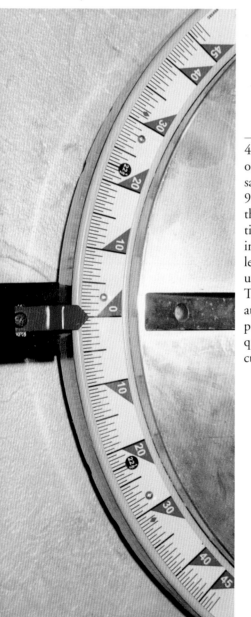

4–11 The scale on power miter saws indicates a 90-degree cut at the 0-degree setting. The markings right and left of center go up to 45 degrees. The scale has automatic stop positions on frequently used cutting angles.

is 90 degrees, you have to subtract the angle wanted, 60 degrees, from 90 degrees to get the setting on the scale, 30 degrees (4–12). Other miter saws have the angles left and right of the 0 point marked, so you do not have to do this subtraction (4–13).

The degree scale has positive stops for several frequently used angles. Typically, these are 0, 22½, 31.6, and 45 degrees right and left. The procedure and controls for setting angles varies a

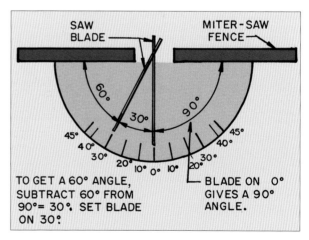

4–12 The miter-saw scale does not indicate the angle actually cut, except for 45 degrees. To set angles with a miter saw, subtract the angle you want from 90 degrees to get the setting for the angle. For example, set the saw on 30 degrees to get a 60-degree angle.

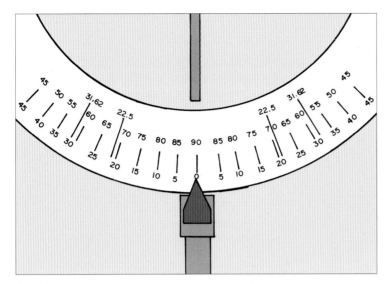

4–13 The scale on some miter saws shows the angles directly.

Once the setting is locked in place, put the casing against the fence, turn on the power, let the blade reach full speed, and lower the saw into the work (4–15). Keep the hand holding the casing well away from the cutting area. Wear eye protection. When the cut is finished, raise the saw, turn it off, and let the blade stop rotating. Always keep the guard over the saw blade. **Do not reach in and try to remove wood pieces while the blade is rotating.**

Another dangerous situation arises when the casing is bowed and you try to cut it. Badly bowed stock should be discarded. If it is necessary to cut a bowed piece, place the bow against the fence (4–16). If the bow is away from the fence, the stock will pinch the saw, causing it to kick back the material.

little from one manufacturer to another. It is important to study the instruction manual that comes with your saw. When you set the saw on the desired angle, be certain to lock it tightly in that position.

The angle of cut can also be set by first establishing the angle with a T-bevel. Then, place the T-bevel against the fence and carefully move the blade next to it (4–14). **Disconnect the machine from the power source before making this setting.**

4–15 When making a cut, wear eye protection, be certain the guard is over the blade, and keep your hands well away from the line of cut.

4–14 The angle of cut can be set using a T-bevel set at the desired number of degrees.

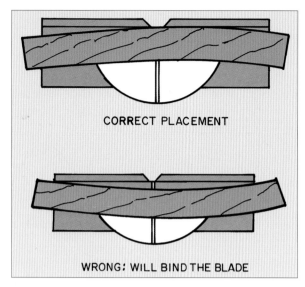

CORRECT PLACEMENT

WRONG: WILL BIND THE BLADE

4–16 If the stock is bowed, place the crown of the bow against the fence of the miter saw. **Courtesy DeWalt Industrial Tool Company**

MITER SAW SAFETY RULES

Detailed safety information for this and other power tools is available from the Power Tool Institute, 1300 Summer Avenue, Cleveland, Ohio 44115-2851. The following are safety rules to observe when using a miter saw:

1. Study the instruction manual provided with the machine.
2. Place the saw away from the path where other workers will be walking.
3. Never use a rip blade or a blade that has been cracked or damaged. Use only sharp crosscutting blades.
4. Keep your hands well away from the blade at all times.
5. After completing a cut, let the blade stop rotating before raising it up from the work. Most saws have an automatic electric brake to stop rotation rapidly.
6. Do not remove small pieces while the blade is rotating.
7. Never try to cut small pieces of wood.
8. Support the work so it rests firmly on the table and against the fence. Get someone to help hold wide or long pieces. Do not try to cut stock freehand.
9. Always disconnect the machine from the power source before you make adjustments to the blade or fence.
10. Do not begin to cut until the motor has reached full speed.
11. When leaving the machine unattended, lock the switch to prevent unauthorized use.

USING A HAND MITER BOX

If you do not have a lot of mitering to do, a hand miter box (4–17) will serve quite well. A hand miter box has a scale just like that described for the power miter saw. As you set the angle, the saw moves along with the adjustment lever. Place the stock against the fence and hold it tight with one hand. Move the fine-toothed saw across the stock, but do not press down or try to force it to cut. The best cut occurs when it goes at its own rate. Very small, inexpensive miter boxes use a backsaw. They are fine for small moldings where great accuracy is not important (4–18).

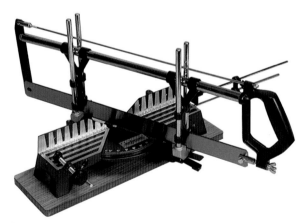

4–17 This manually operated miter box will accurately cut miters and other angles right and left of the 0-degree scale setting. **Courtesy Adjustable Clamp Company**

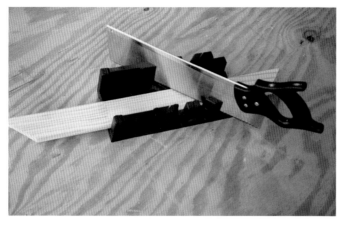

4–18 This small, inexpensive plastic miter box uses a small backsaw. It is not adjustable, but will cut 90- and 45-degree right and left angles, and a 45-degree bevel.

ADJUSTING MITERS

Miters will not always fit tightly on the first try, and the casing will have to be removed and the miters adjusted. Illus. **4–19** shows the types of corrections that may need to be made. The top drawing shows a miter that has no gaps and is square. This is how all miters should appear, but, because of variances in the wall surface or the window frame, this will not always be the case. The casing must always keep the reveal uniform on all sides of the window.

Illus. **4–19** also shows a miter touching at the toe but open at the heel, and a miter touching at the heel but open at the toe. In these drawings, the amount of gap is greatly exaggerated. In actual practice, the gap will be much smaller. Usually the gap is in the range of ⅟₆₄ to ⅟₃₂ inch.

CLOSING GAPS IN MITERS

The gap in a miter may be closed by removing wood from one side of the miter. This may be done several ways. One way is to place the casing on the miter saw, place a small wedge between the casing and the fence, and make a very fine cut across the miter. All you will get will be some

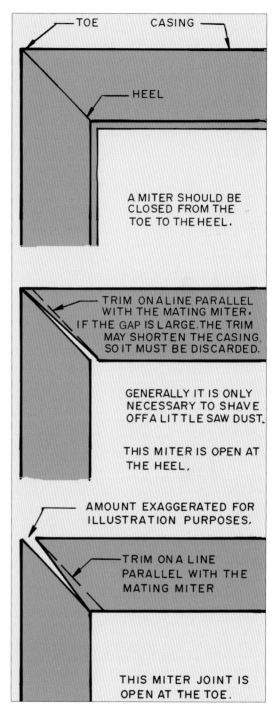

4–19 Typical problems that can occur as you try to install a mitered casing.

sawdust off the end you are lowering (**4–20**). The amount to remove is determined by examining the joint and observing the size of the gap. Usually, a thin piece of cardboard gives enough change in the angle of the miter to close the joint. Check the joint after cutting, and remove more if needed.

Another way to lightly trim a miter is to use a block plane, as shown in **4–21**. Plane "down the slope." Slant the plane a little across the surface, and move it with a slicing motion (**4–22**). Take very light cuts, and be certain the plane is very sharp. Caution must be exercised to never slant the surface to the front; however, it can be slant-ed a little to the back surface. This will actually help close the miter. Often, a miter will have parallel edges, but will not close. This is an indication that the surfaces of the miter slant toward the front. Remove some wood off the back edge until the miter closes, as shown in **4–23**.

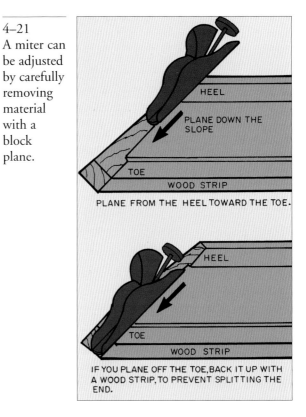

4–21 A miter can be adjusted by carefully removing material with a block plane.

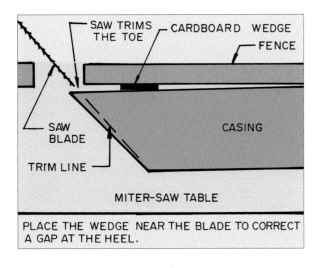

PLACE THE WEDGE NEAR THE BLADE TO CORRECT A GAP AT THE HEEL.

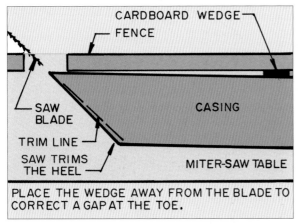

PLACE THE WEDGE AWAY FROM THE BLADE TO CORRECT A GAP AT THE TOE.

4–20 A miter can be adjusted by trimming a bit of sawdust off the toe or heel as needed to make it close.

4–22 When lightly adjusting a miter, place it on a support and plane on a slight angle across the surface of the miter.

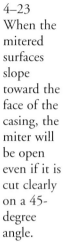

4–23 When the mitered surfaces slope toward the face of the casing, the miter will be open even if it is cut clearly on a 45-degree angle.

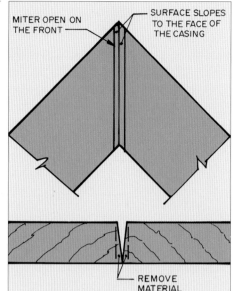

MITER OPEN ON THE FRONT

SURFACE SLOPES TO THE FACE OF THE CASING

REMOVE MATERIAL

INSTALLING PICTURE-FRAME MITERED CASING

Picture-frame is casing installed on all four sides with the corner usually mitered (**4–24**).

Begin by marking the **reveal** on the corners of the casing. A reveal is the amount the casing is set back from the edge of the frame. Typically, it is either 3/16 or 1/4 inch. An easy way to mark it off is to set a combination square with the blade extending out the amount of the reveal (**4–25**). Make a pencil mark on each corner and several between the corners (**4–26**).

Picture-frame miter casing has two lengths of casing. The length of each can be measured on the short or long side of the casing. In **4–27**, the lengths are taken on the short side of the casing.

4–24 This vinyl bow window is attractively cased with stained wood picture-frame casing. **Courtesy Chelsea Building Products, Inc.**

This includes the distance inside the jambs plus twice the reveal. When laying out the head casing, mark the points of intersection of the reveal lines. Then lightly mark a pencil line indicating the direction of cut (**4–28**). Do not press so hard as to score the surface of the casing. This will help you cut in the proper direction when you get to the miter saw.

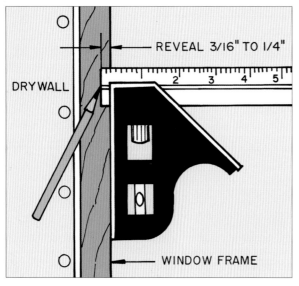

4–25 A reveal, the amount the casing is set back from the edge of the frame, can be marked using a combination square. It is typically 3/16 inch.

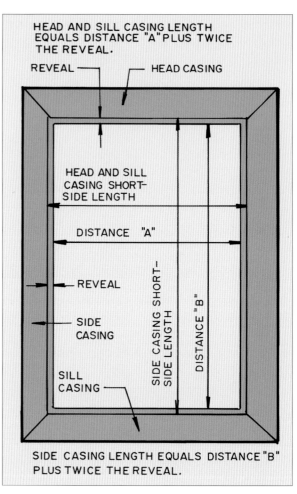

4–27 The short side of the casing is equal to the distance between the jambs plus twice the reveal.

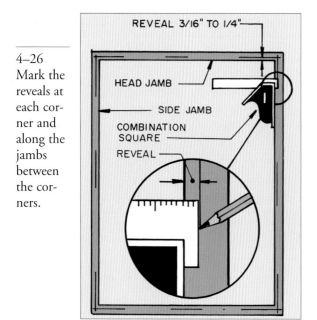

4–26 Mark the reveals at each corner and along the jambs between the corners.

4–28 To lay out the head casing on the short side, mark the points of intersection at each corner. Lightly sketch a line showing the direction of the cut.

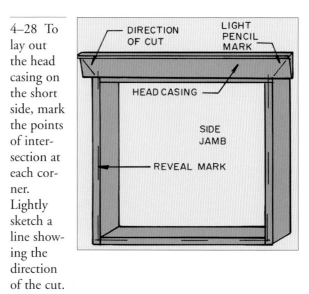

Interior finish carpenters approach the installation of mitered casing in different ways. The following is one suggested procedure: Start the installation by nailing the head casing to the jamb edge with 3d or 4d finishing nails. Do not set the nails too firmly, because it may be necessary to adjust the casing in order to get a tightly closed miter. Be certain the casing fits flat against the surface of the drywall. If it does not, it may be necessary to file off some of the drywall or plane a little off the jamb. (See **5–8 to 5–11** in Chapter 5.) Be certain the casing lines up with the reveal marks, just covering them (**4–29**).

Now install the side casings, being certain that they line up with the reveal marks and that the miters close. Nail them to the frame. Check again to be certain the miters are closed after nailing. If the miters do not close, adjust them as described in the next paragraph.

Finally, cut and install the sill casing. You can measure the actual length required, or cut a miter on one end, place the stock across the bottom, and mark the location of the other end. Cut and install the bottom casing. If all miters are closed, finish nailing by placing nails about 8 to 10 inches apart along the edge of the frame. Then, using 6d or 8d finishing nails, nail the thick side of the casing to the header or stud. Suggested nailing patterns are shown in **4–30**. When installing hardwood casing, it may be necessary to drill a small hole for each nail. The hole should be a little smaller than the diameter of the nail.

The use of a power nailer greatly speeds up this type of work. It also leaves one hand free to hold the casing in place (**4–31**), and is also less likely to split the casing.

Instead of nailing the miter as shown in **4–30**, you can install a plate. A plate joiner is shown in Chapter 3, and a typical installation is shown in Chapter 6.

Some finish carpenters use files to correct miters. While files do remove material, it is dif-

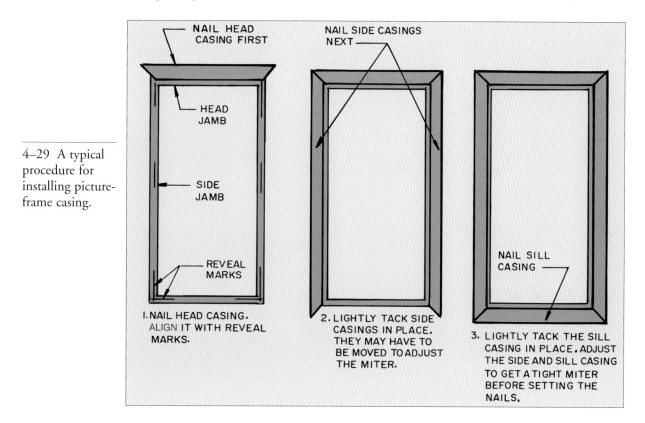

4–29 A typical procedure for installing picture-frame casing.

ficult to keep a surface flat or straight with one. Filing tends to round the surface, which, will prevent the miter from closing. Wood chisels, multi-blade forming tools, or a block plane should be used instead.

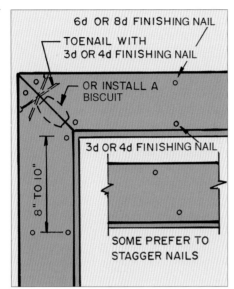

4–30 A suggested nailing pattern for installing casing. Some prefer to stagger the nails. Notice how the miters can be fastened.

4–31 Power nailers speed up the installation of the casing and will not damage the surface. They also leave one hand free to hold the casing. Courtesy Paslode, an Illinois Tool Works company

INSTALLING THE STOOL & APRON

Windows can also be trimmed using a stool and apron at the sill. Typical stool profiles are shown in 4–32. They are available as stock material from building-material suppliers. In some cases, window manufacturers have stool material available for their specific window units. A stool with an angled bottom surface is used on windows made with a sloping sill (4–33).

FITTING THE STOOL

First, cut the stool to its finish length. This includes the width between the window side jambs, plus two times the reveal, plus two times the width of the casing, plus two times the overhang. The horn is equal to the width of the casing plus the reveal and any extension allowed beyond the casing. This is typically ½ to 1 inch.

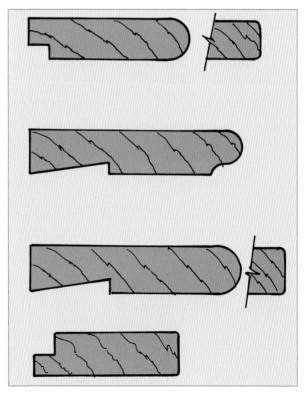

4–32 Typical stock stools.

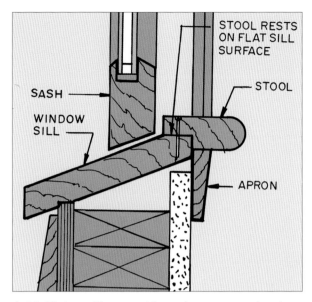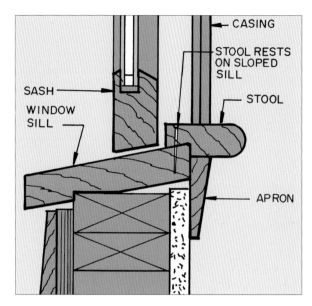

4–33 Various sill types with stools, aprons, and casing.

It is simply decorative, so it can be any distance desired (4–34).

Next, measure the window unit to see how deep a notch is needed to form each horn; lay this out on the stool. Cut the horn notches, and fit the stool into the window frame. Trim and adjust it, as necessary, to get the required fit (4–35). Be certain the stool is level. If it is not, it may be necessary to shim it. The stool must be ¹⁄₁₆ to ⅛ inch clear of the sash.

If the window needs a jamb extension, some carpenters prefer to install the stool first and then add the jamb extension. Others prefer to add the jamb extension and cut the stool to fit it (4–36).

When you are satisfied with the fit of the stool, consider shaping the ends of the horn (4–37). They can be left square, as cut, if you sand out the saw marks. However, some finish carpenters prefer to round the horn or shape a

4–34 A layout showing how to find the length of the stool, apron, and horn.

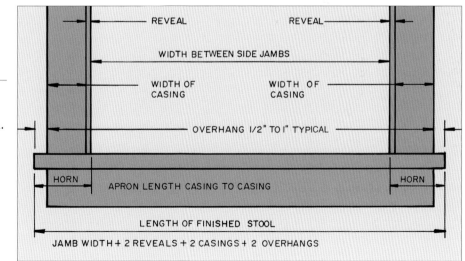

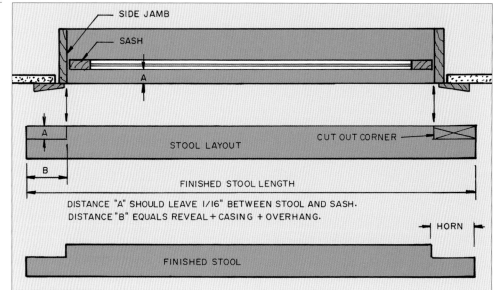

4–35 Typical layout for a window stool. This may vary a bit, depending upon the window being used.

SIDE JAMB

SASH

A

STOOL LAYOUT

CUT OUT CORNER

A

B

FINISHED STOOL LENGTH

DISTANCE "A" SHOULD LEAVE 1/16" BETWEEN STOOL AND SASH.
DISTANCE "B" EQUALS REVEAL + CASING + OVERHANG.

HORN

FINISHED STOOL

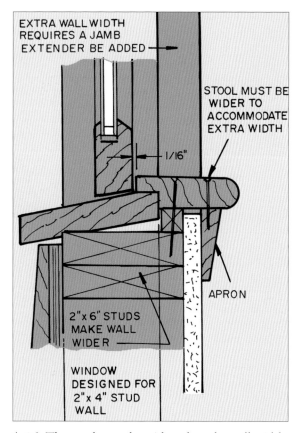

EXTRA WALL WIDTH REQUIRES A JAMB EXTENDER BE ADDED

STOOL MUST BE WIDER TO ACCOMMODATE EXTRA WIDTH

1/16"

APRON

2"x 6" STUDS MAKE WALL WIDER

WINDOW DESIGNED FOR 2"x 4" STUD WALL

4–36 The stool must be wider when the wall width requires that jamb extenders be added. Nail the stool to the sill and apron. When possible, nail into the rough two-inch sill framing below.

profile. The stool with the mitered return provides face grain on the exposed end. (A mitered return is shown in 4–37.) The other two ends expose end grain. Face grain is especially nice when the stool is a hardwood and a smooth return surface is desired.

Now nail the stool in place. Nail through the stool into the rough sill (if possible) using 8d finishing nails. After the apron is installed, nail the stool to it with 6d finishing nails (4–36). Since the actual design of window frames varies, the stool may have to be nailed into the sill below it. On some windows, it is not possible to install a stool.

INSTALLING THE APRON

The apron is usually made from the same material used to trim the window. It is placed below the stool and covers any opening between the sill and the drywall. It is placed with the thick edge up against the stool (4–36 and 4–37).

The length of the apron is usually made the same as the overall distance from one side of the window casing to the other. This permits the stool horn to extend beyond the apron (4–34). Some people prefer to extend the apron a little beyond the casing.

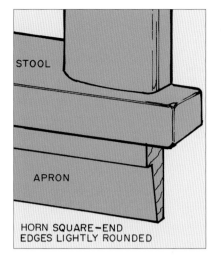

4–37 The horn can be shaped to provide the appearance desired.

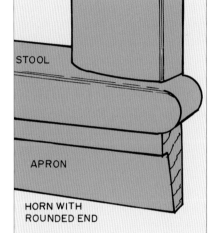

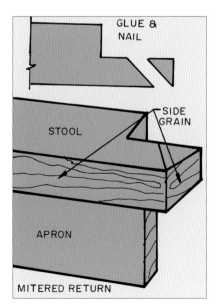

The ends of the apron may be finished in several ways. The easiest is to cut them square (4–38). Another is to cut the profile of the face design on the end (4–39). A third way is to miter the ends and glue in a return (4–40).

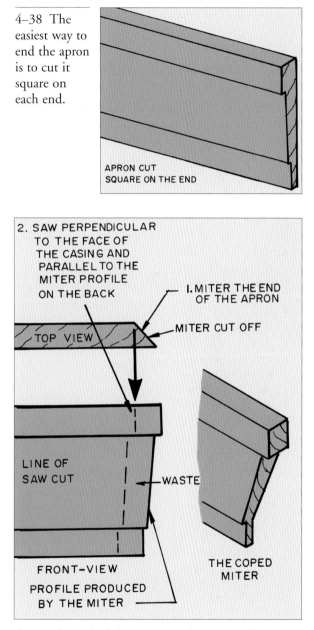

4–38 The easiest way to end the apron is to cut it square on each end.

4–39 The end of the apron can be cut to reflect the profile of the casing.

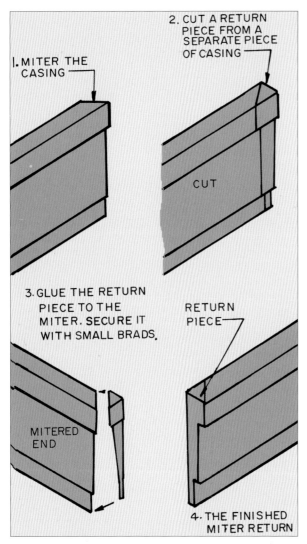

1. MITER THE CASING

2. CUT A RETURN PIECE FROM A SEPARATE PIECE OF CASING

CUT

3. GLUE THE RETURN PIECE TO THE MITER. SECURE IT WITH SMALL BRADS.

RETURN PIECE

MITERED END

4. THE FINISHED MITER RETURN

4–40 When hardwood casing is used, many prefer to miter the end of the apron.

4–41 Mitered casing and a stool and apron are popular ways to case window openings. **Courtesy Chelsea Building Products, Inc.**

INSTALLING MITERED CASING WITH A STOOL & APRON

Mitered window casing with a stool and apron is shown in **4–41**. Begin by installing the stool as described earlier. The top and side casings are to be installed as described for picture-frame casing; however, the side casing should be cut square on the bottom where it butts the stool.

After the stool is installed, cut and install the head casing, lining it up with the reveal marks.

Then measure the length of each side casing. Measure both sides, because there is often a slight difference in their lengths. Some measure the short side, while others measure the long side. Place the side casing on the reveal marks (**4–42**). Check the miter and adjust it if it is not tight. Check the bottom to be certain it is square with the stool. Nail the casing and the corners as shown in **4–30**.

INSTALLING WINDOW CASING WITH MULLIONS

Often, windows are installed in a series with a mullion[3] between them, as shown in **4–43**. Frequently, the mullion is too narrow to permit the installation of two full-width casings. Methods of installing windows with mullions vary. Manufacturers have recommended installa-

[3]Mullions are slender vertical members separating windows.

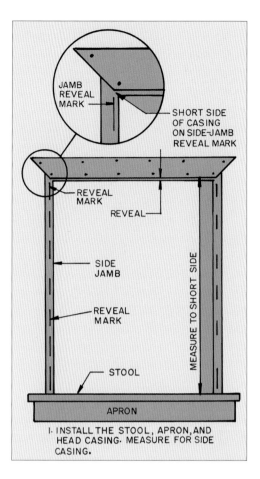

JAMB REVEAL MARK

SHORT SIDE OF CASING ON SIDE-JAMB REVEAL MARK

REVEAL MARK

REVEAL

SIDE JAMB

MEASURE TO SHORT SIDE

REVEAL MARK

STOOL

APRON

1. INSTALL THE STOOL, APRON, AND HEAD CASING. MEASURE FOR SIDE CASING.

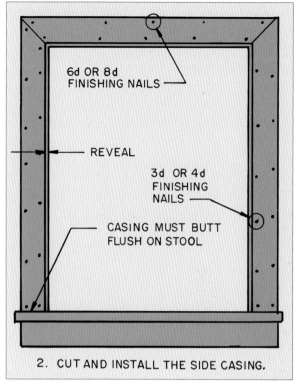

6d OR 8d FINISHING NAILS

REVEAL

3d OR 4d FINISHING NAILS

CASING MUST BUTT FLUSH ON STOOL

2. CUT AND INSTALL THE SIDE CASING.

4–42 Install the stool and apron before the head and side casings.

tion procedures for their products. Some typical details are shown in 4–44 to 4–46.

If the mullion is narrow, it is not possible to install casing that matches that used on the head and side jambs. One solution is to cover the mullion with a thin material such as ¼-inch plywood. If the casing is hardwood, select plywood with a hardwood veneer that can be finished in the same manner as the casing. If you are adding a stool, install it and the head and side casing before the mullion casing. If

4–43 This multiple-window installation has rather wide mullions between each window. The width used depends upon the appearance desired. Courtesy Weather Shield Manufacturing Company

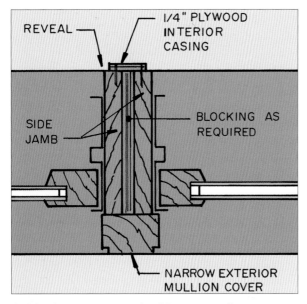

4–44 This narrow, non-load-bearing mullion is formed by butting the window frames and using a little blocking if needed.

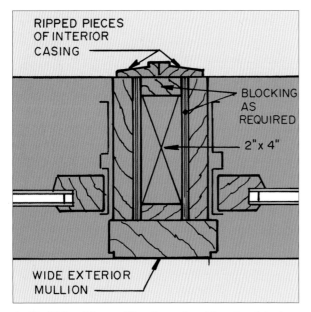

4–45 This wider mullion is made with a 2 x 4-inch stud spacer. It will carry some load. Blocking can be used to vary the width.

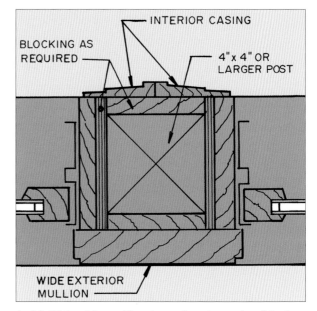

4–46 This wide mullion is made using a 4 × 4-inch post. Any size mullion can be made by varying the internal load-bearing members and the blocking.

you are not adding a stool, install the entire picture-frame casing and then the mullion casing. Be certain in both cases to mark the reveal on the jambs as you did for the regular casing.

Wider mullions can also be covered with a thin casing, such as ¼-inch plywood. However, on many jobs the same casing that is used on the head and side jambs is used on the mullion. Often, the width of the mullion will not permit the use of the full-width casing, so the two pieces of casing must be ripped narrower. Measure the distance between the reveal marks and divide this in half. Cut the casing to width, removing the thicker edge. Then miter the top end (4–47).

The head casing pieces are mitered at the reveal mark on the mullion. Then the tip is cut off at a point equal to one-half the width between the reveals.

Installation takes a bit of adjustment. First, temporarily tack the head casing along the reveal mark. Then take both pieces of mullion casing and put them in place as if they were one piece. Now, adjust them until they are on their reveal

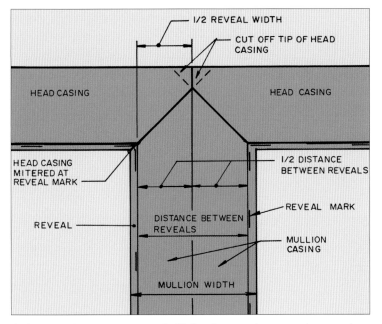

4–47 Cut the casing so it is half the distance between the reveals, and miter the top end. Trim the tip off the head casing and tack the head casing to the head jamb.

marks and the miter at the head casing is closed. When all appears to be set, tack them to the jambs. Loosen and retack them as necessary. Finally, nail them to the jambs and mullion with 6d or 8d finishing nails as needed (4–48).

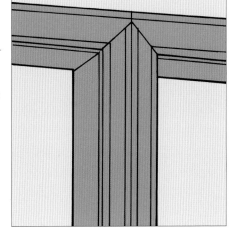

4–48 The finished intersection between the head and mullion casings.

INSTALLING BUTTED CASING WITH A STOOL

Butted window casing is installed by first cutting the side casings to length. Both ends are cut square. The top end should touch the reveal marks. Be certain that the end butting the stool has a tight, closed joint when the casing is in line with the reveal marks.

The head casing may be cut so it is flush with the outside edges of the side casings or cut the same length as the stool, providing a small reveal (4–49). However, the head casing may be even wider and have various moldings applied to it. Often, the door and window head casings in a room are of the same design.

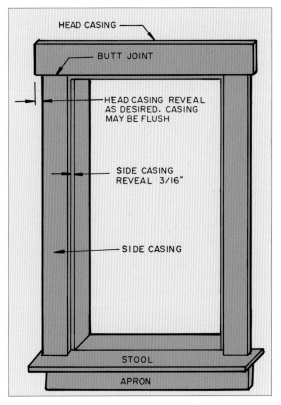

4–49 Flat casing can be installed with a reveal, or the side casing can be flush with the end of the head casing.

USING ROSETTES

Butted casings on some traditional houses use rosettes. The rosettes, also called corner blocks, are available from building-material suppliers. They are typically wider than the casing, and overhang it the same amount as the casing reveal (4–50).

Begin by installing the sill and apron as described earlier. Then install the side casings in the manner described for butted casings. Next, nail the rosettes on top of the side casings. Some prefer to align one side of the rosette with the edge of the casing, while others align it with the window jamb (4–51). Then measure the distance between the rosettes and cut the head casing. It must butt tightly against each rosette. This will usually take some minor adjustments. It helps if the butting ends can have a slight bevel toward the back of the casing (4–52). This makes it easier to get a tight fit.

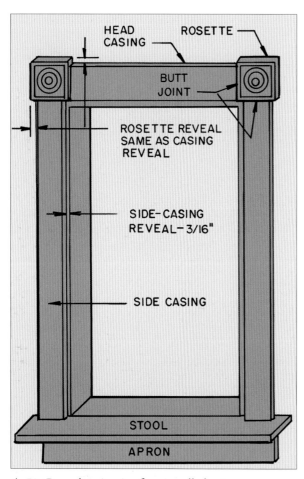

4–50 Butted casing is often installed using rosettes.

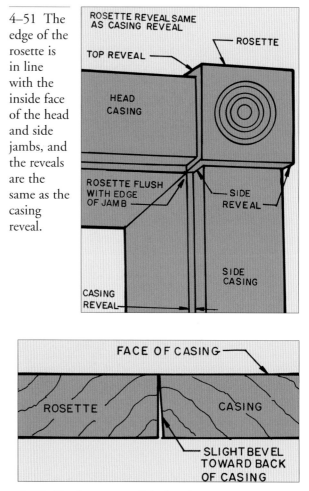

4–51 The edge of the rosette is in line with the inside face of the head and side jambs, and the reveals are the same as the casing reveal.

4–52 Cutting a very slight bevel on one or two butting pieces helps produce a tightly closed joint.

INSTALLING CASING ON VINYL WINDOWS

The exact design of vinyl windows varies, but the following is typical. Vinyl windows are installed with a nailing fin. This leaves some of the rough opening exposed. This can be covered with wood trim, as shown in **4–53**, or with drywall. After the jamb-extension trim has been installed, the casing is placed in the manner described earlier.

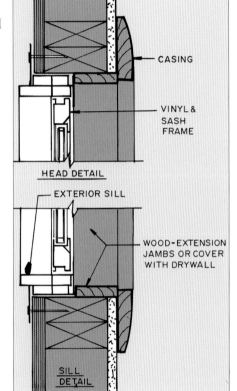

4–53 When vinyl windows are installed, the interior of the rough window opening is trimmed with wood extension jambs and the casing is installed in the normal manner.

CASING

VINYL & SASH FRAME

HEAD DETAIL

EXTERIOR SILL

WOOD-EXTENSION JAMBS OR COVER WITH DRYWALL

SILL DETAIL

CIRCULAR & ELLIPTICAL HEAD CASINGS

The manufacturers of circle-top windows also supply the necessary casing to fit the unit (**4–54**). Usually several casing styles are available.

SPECIAL HEAD-CASING DESIGNS

The designer may copy an ornate traditional head casing or prepare an original design. Each situation may have particular requirements so that detailed drawings are necessary to allow the finish carpenter to build exactly what you want. Special casings can also be built at a cabinetmaking or millwork shop, and delivered to the job site already assembled. Details for installation are as described in the earlier parts of this chapter.

4–54 Circle-top windows have casing manufactured to fit the curved jamb. **Courtesy Weather Shield Windows and Doors**

CHAPTER 5

Door Casing

The door and casing make a major design statement in the overall appearance of a room. The use of sidelights and a transom provide additional features and often require additional trim. The installation in **5–1** has stained wood doors and casings which set off the door against the lighter colored walls, wainscoting, and baseboard. Often, the casing is painted the same color as the walls and is not a major visible feature. In **5–2**, it blends with the walls and wainscoting. The profile provides shadow lines that enhance the overall appearance, but do not dominate the wall. Here, the doors are the dominant feature.

If you want something more dramatic, install a unit with sidelights, a crown-molding head trim, and a circle transom, as shown in **5–3**. Notice the great details, such as the plinth blocks at the floor and the keystone[1] inserted at the top of the circle-top casing.

Another installation, shown in **5–4**, uses a custom-built door in a residence that is a timber-framed building. The door and flat trim reflect the more rustic interior typical of this type of house. The cleats across the boards that form the door were stained darker, providing an interesting contrast.

Another design approach is to install acrylic blocks as sidelights (**5–5**). This lets in natural light, yet maintains privacy. Often, a wide mullion is placed between the door framing and the sidelights. The casing must be designed to cover this mullion, and follow the style of the other door and window casings in the room.

Many homes use what some refer to as a "bare-minimum casing" (**5–6**). This is stock molding that is usually around 2½ inches wide. It does make a neat framing and, with the use of color, can provide a nice effect.

Extra-wide, flat casing creates a heavy, massive appearance (**5–7**). It gives greater prominence to

5–1 This wood door with sidelights and a transom was stained, enhancing the natural grain of the wood.
Courtesy Simpson Door Company

a rather narrow door. Remember, door casing can be made up of several pieces of molding, producing a very sculptural, decorative installation. Also, the use of plinth blocks (**5–8**), fluted casing (**5–9**), and built-up decorative head casing (**5–10**) all add to your choices.

Another great feature for classical homes is a pediment over the main entrance door (**5–11**). Another very attractive installation uses a wide casing with returns that form a narrow, flat column. Stock molding is mitered around it, and some form of a multipiece header is built to cross the door opening (**5–12**).

---

[1]*A keystone is the wedge-shaped piece at the crown of the door arch that locks the other pieces in place.*

5–2 These solid vinyl sliding patio doors are the major feature of the outside wall. The casing and wainscoting blend together, providing a comfortable background. **Courtesy Simonton Windows**

5–4 This custom-built door with flat trim complements the rustic, timber-framed home and reflects those doors typical of early homes built in rural America. Doors of this type are often referred to as board-and-batten doors. **Courtesy Dreaming Creek Timber Frame Homes, Inc.**

5–3 This dramatic, elaborately trimmed entryway provides a major focal point in the room or foyer. **Courtesy Simonton Windows**

5–5 This entrance uses acrylic blocks for the sidelights. Notice the wide mullions, which require wide casing that is compatible with that used on the windows and interior doors. **Courtesy Hy-Lite Products, Inc.**

5–6 Minimum-size stock casing will provide an adequate finish, but does not add much to the décor.

5–9 Fluted casing is very decorative and typical of European neoclassicism of the late 18th and 19th centuries. Notice that the casing terminates on plinths at the floor. **Courtesy Ornamental Moldings**

5–7 The beauty of this wide, flat casing enhances the stained wood door and makes it appear wider.

5–10 Decorative head casing has been used in most architectural periods. The design varies with the practices in vogue at the time. Notice the fluted side casing. **Courtesy White River Hardwoods Woodworks, Inc. and The Hardwood Council**

5–8 Plinth blocks provide a base where the casing terminates at the floor. **Courtesy Simonton Windows**

5–11 A pediment makes a significant architectural statement. **Courtesy Architectural Products by Outwater, LLC (800-835-4400)**

5–12 This wide, fluted flat column was used to case the door opening, giving it additional depth. Stock molding has been mitered at the top and a multipiece header spans the door opening.

CHECKING DOOR INSTALLATIONS

If you installed the doors yourself, you know whether they have been set level and plumb in the rough opening. If they are already in place when you arrive to install the casing and any special head cornices, check the frame with a level to see if the side jambs are plumb and the head jamb is level. If they are not, it will affect the angles on the miters or the square cuts against a plinth or rosette.

Check to see if the drywall extends beyond the doorjamb (**5–13**). If it extends a small amount, the surface of the drywall can be planed or sanded on a slight angle until it is flush with the edge of the frame, the casing will fit flat on the wall, and the miter joint will close (**5–14**). A surform tool is good for this job.

Another situation occurs if the doorjamb extends beyond the face of the drywall (**5–15**). A small amount can be corrected by planing the edge of the door frame flush with the drywall (**5–16**).

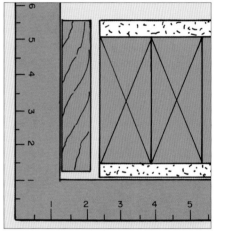

5–13 Check to see if the drywall is flush with the edge of the door frame. If it extends beyond it, dress it down with a surform plane.

5–14 The drywall that extends beyond the edge of the doorjamb can be feathered out with a surform plane so the casing will fit flat against it.

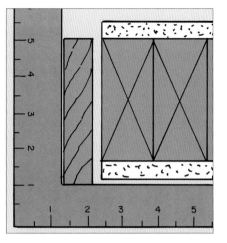

5–15 Check to see if the edge of the doorjamb extends beyond the surface of the drywall.

5–16 When the doorjamb extends beyond the edge of the drywall, plane the jamb flush.

DOOR CASING STYLES

Door casing moldings are available in a wide variety of sizes and profiles. Examples of typical stock casing are given in Chapter 1. The design and components can vary, which influences the installation procedure.

The most common technique is to use some form of machined casing and join it at the top corners with a miter (5–17). Some classical styles require the use of plinths and rosettes (5–18). The casing is often heavily fluted. The various parts meet with butt joints.

Various compound head casings can be constructed using standard moldings (5–19). Typically, the side casings butt the material,

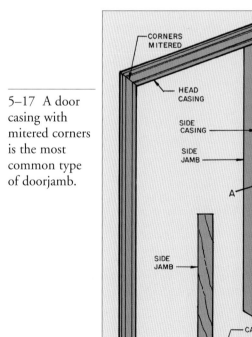
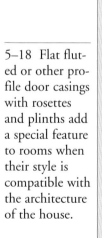

5–17 A door casing with mitered corners is the most common type of doorjamb.

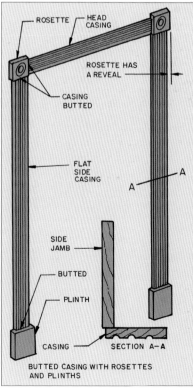

5–18 Flat fluted or other profile door casings with rosettes and plinths add a special feature to rooms when their style is compatible with the architecture of the house.

forming the head casing. A very simple installation, made with flat casing on the side jambs and head jamb, uses a butt joint (**5–20**). It can be enhanced by adding a small molding as a base cap or nosing. The head casing can extend beyond the side casings, providing a small reveal.

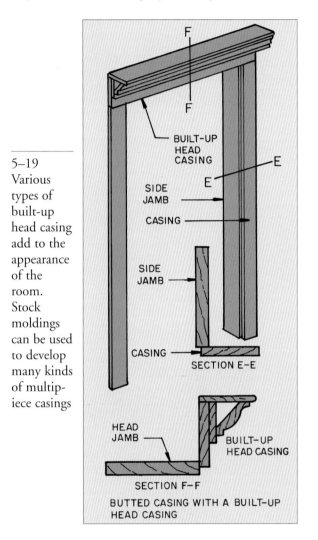

5–19 Various types of built-up head casing add to the appearance of the room. Stock moldings can be used to develop many kinds of multipiece casings

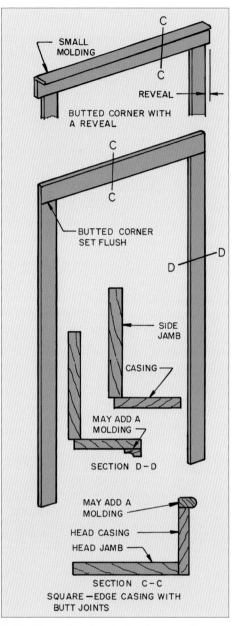

5–20 Butted casing is the simplest form used and is easy to install.

INSTALLING DOOR CASING

When the job has doors that are not prehung, or does not use precut frame and casing kits, it is necessary to make these from stock lengths of casing molding. This requires careful measuring, accurate cutting, and protection of the casing from damage as it is processed.

First, check the conditions between the jamb and the drywall. Make any corrections necessary so the casing will lie flat on the wall. Then follow the directions that follow.

INSTALLING MITERED CASING

Begin by marking the reveals on the head and side jambs. The easiest way to do this is with a combination square (**5–21**). The reveal most commonly used is ³⁄₁₆ to ¼ inch. Additional details are given in Chapter 4.

While finish carpenters use several different procedures to install mitered casing, the following is typical: First, cut and install the head casing. Measure from the intersection of the reveal marks on each side (**5–22**). Cut the miters on each end. Place the casing on the reveal marks and tack it to the jamb. Use 3d or 4d (1¼- or 1½-inch) finishing nails. Leave the nails sticking out a bit so they can be pulled if it is necessary to move the casing. Place nails about every 12 inches. When using hardwood casing, pilot holes will most likely need to be drilled for each nail.

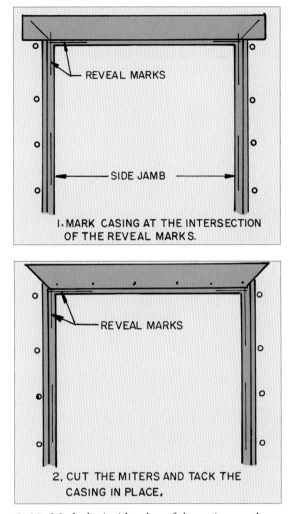

1. MARK CASING AT THE INTERSECTION OF THE REVEAL MARKS.

2. CUT THE MITERS AND TACK THE CASING IN PLACE.

5–22 Mark the inside edge of the casing on the intersection of the reveal marks. Cut the miters and install the casing at the inside point of this intersection.

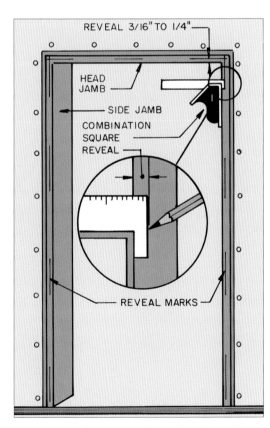

5–21 Mark the reveals along the side and head jambs.

Next, place a piece of casing along the side jambs and mark the corner of the reveal and the angle of cut on it. Cut the miters and place the casing against the jamb along the reveal marks (**5–23**). Check the miters. If they are tight, nail the side casings to the jamb and stud (**5–24**).

Some people place glue in the miter joint before nailing the casing in place. Whether or not you use glue, the next step is to nail the miter as shown in **5–25**. This pulls the miter together and helps keep it closed over the years.

5–23 Place the casing along the side jambs, mark and cut the miters, and install the casing along the reveal marks.

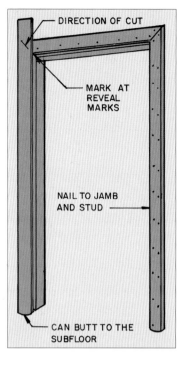

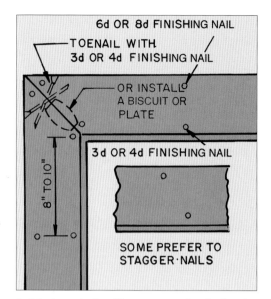

5–25 A typical nailing pattern for the head and side casings.

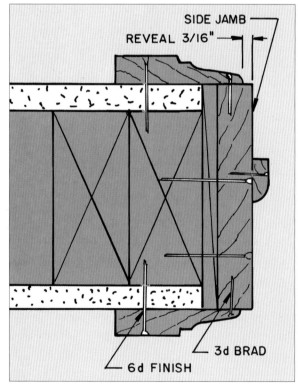

5–24 The side casings are nailed to the jamb and the trimmer stud.

A plate or biscuit can be installed instead that, when glued, holds the miter tightly closed. Information on this installation is given in Chapters 1 and 6.

Since the miter will often not close tightly, some adjustments must be made. These techniques are discussed in Chapter 4.

Once the miters are closed and the casings are tacked in place, drive the nails close to the surface and set them below the surface of the casings (5–26). The painter will fill these holes and sand the surface in preparation for the final finish. See Chapter 9 for information on finishing.

Often, it is necessary to trim the bottom of the side casings so the flooring (made of material such as wood, laminate, or vinyl) can slide under them. If you know the thickness of the floor, this amount can be trimmed off before the casing is installed. It is often necessary to trim it later. To do this, place a piece of the flooring next to the casing and trim it with a special saw as shown in 5–27.

The previous discussion relates to hand-nailing the casing to the jamb and trimmer stud. A power nailer will greatly speed up the work and leaves one's hands free to hold the casing as it is nailed. Use 1½-inch slight-headed finishing nails (**5–28**).

See Chapter 4 for information on laying out and cutting miters.

5–26 The finishing nails are set below the surface of the casings with a nail set. **Courtesy Stanley Tools**

5–27 Place a piece of the flooring to be used next to the door casing and trim it off with this special saw so it will slide under the casing when the floor is laid. **Courtesy Robert Bosch Tool Corporation**

CUTTING & ADJUSTING MITERS

Miters on casing are most accurately cut with a power miter saw. Sometimes, due to slight irregularities in the frame installation or drywall, the miter will not close tightly and have to be trimmed. This procedure, and detailed information on cutting miters, is given in Chapter 4.

5–28 It is much faster to secure interior moldings if you use a power nailer. **Courtesy Ornamental Moldings**

INSTALLING BUTTED CASING

Mark the reveals on the head and side jambs as shown for mitered casing (**5–21**). Cut the side casings to length. Mark them where the head and side jambs' reveal marks cross. Be certain the casing is square with the floor (**5–29**). If the finished floor has been installed before the door casing, you can get a tighter fit with the floor if you cut the bottom end of the casing on a slight slant toward the back (**5–30**).

After the side casings are nailed to the jamb, the head casing is marked and cut to length. It may be cut flush with the casings, or have a slight reveal (**5–29**). If the head will have moldings attached, it should be assembled before it is installed.

Butted casings are nailed as shown in **5–31**.

Sometimes the head casing is made from stock thicker than the side casings. This produces a decorative reveal at the corner (**5–32**).

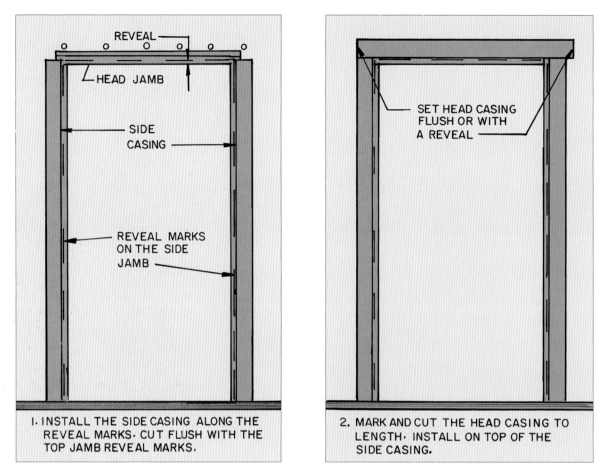

REVEAL	SET HEAD CASING

HEAD JAMB

SIDE CASING

REVEAL MARKS ON THE SIDE JAMB

SET HEAD CASING FLUSH OR WITH A REVEAL

1. INSTALL THE SIDE CASING ALONG THE REVEAL MARKS. CUT FLUSH WITH THE TOP JAMB REVEAL MARKS.

2. MARK AND CUT THE HEAD CASING TO LENGTH. INSTALL ON TOP OF THE SIDE CASING.

5–29 When installing door casing with butted joints, cut the side casings to length and install them. Then cut and install the head casing.

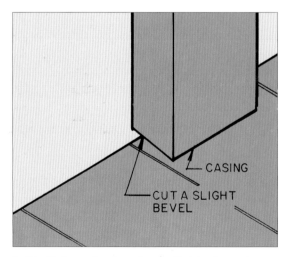

CASING

CUT A SLIGHT BEVEL

5–30 If the casing is to butt a finished wood floor, the bottom edge can be given a slight bevel; this produces a better joint.

Butted casing with plinth and corner blocks is installed much in the same way as described for plain butted casing. The exact sequence of events will vary, but the following is typical: First, install the plinth blocks (5–33). They are installed flush with the edge of the door frame (5–34). They are usually wider than the casing, so they have a reveal on all sides.

Next, cut the side casings to length and install them on the reveal marks on the jamb (5–35). Then install the rosettes. Set them flush with the edge of the head and side jambs (5–36).

Finally, cut and install the head casing between the rosettes. Measure and cut it carefully. You can bevel the square end slightly to get a tighter fit (5–37).

It should be noted that butted casing can be constructed in only one of two ways: with plinth blocks and mitered or butted head casing; or with rosettes and side casings that run to the floor.

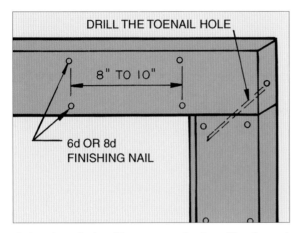

5–31 A typical nailing pattern for installing butted casing.

5–33 The fluted door casing rests on a plinth at the floor. Notice the multipiece cornice on top of the casing that spans the door opening.
Courtesy Simonton Windows

5–32 The appearance of the butted corner can be improved by using a thicker head casing, producing a reveal.

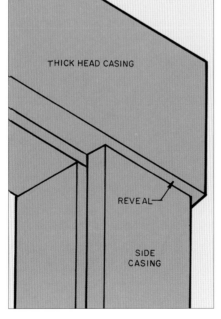

5–34 The plinth is set flush with the edge of the door frame.

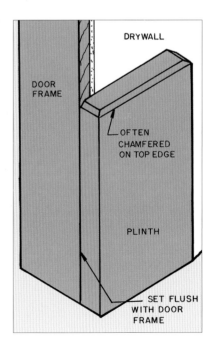

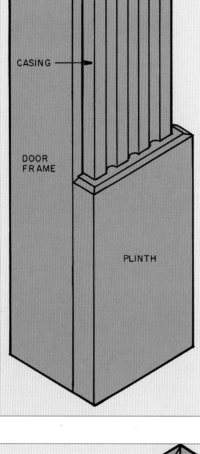

5–35 The plinth is usually wider than the casing, providing a reveal on three sides.

CASING

DOOR FRAME

PLINTH

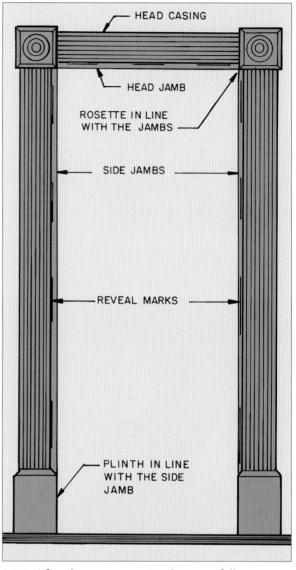

HEAD CASING

HEAD JAMB

ROSETTE IN LINE WITH THE JAMBS

SIDE JAMBS

REVEAL MARKS

PLINTH IN LINE WITH THE SIDE JAMB

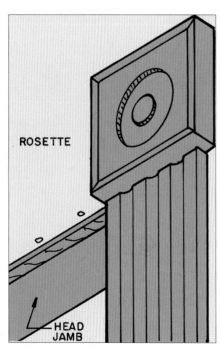

5–36 Install the rosettes on top of the side casings. Line them up with the edges of the doorjamb.

ROSETTE

HEAD JAMB

5–37 After the rosettes are in place, carefully measure, cut, and install the head casing.

CHAPTER 6

Installing Base, Shoe, Chair-Rail & Picture Molding

While the process can vary with different contractors and geographic regions, generally when the trim carpenter arrives on the job the door openings have been framed, the windows have been installed, and the drywall is in place (6–1). Sometimes the interior door frames are installed by the trim carpenter.

Do not move moldings into the house until the drywall compound and texturing on the surface (if any) is thoroughly dry and the humidity in the house is normal. The air temperature in the house should be near normal (65 to 70°F). The trim should be kiln-dried and kept covered to protect it from moisture and possible damage.

6–1 This trim is being installed after the unfinished hardwood flooring has been laid. Heavy cardboard mats have been placed down to protect the floor. The manually operated miter saw is lightweight, easily supported by wood sawhorses, and easily moved around the house. **Courtesy Stanley Tools**

WHEN SHOULD BASEBOARD BE INSTALLED?

Should baseboard be installed before or after the flooring? This decision is sometimes made by flooring and trim carpenters; however, it generally depends upon the type of flooring to be laid.

Many prefer to have **unfinished hardwood flooring** laid before installing the trim (6–2). This way, they can do the preliminary sanding up to the wall without damaging the baseboard.

6–2 This unfinished hardwood flooring is laid over red rosin paper. When completely installed, it will be swept clean and ready for the trim carpenter to begin work on.

The baseboard can then be installed, but care must be taken to prevent damage to the unfinished wood flooring. If **prefinished wood flooring** is to be used and has been installed first, many people lay protective red rosin paper over the area in which the baseboard is being laid and over all areas where traffic is expected. Generally, prefinished wood flooring is laid after the baseboard is in place (6–3). Care must be taken to

6–3 This prefinished hardwood flooring is being laid after the trim has been installed and the walls and trim painted.

protect the finished floor while the shoe molding is being installed.

Many prefer to have **vinyl floor covering** installed after the trim is in place. It is often installed by stapling it along the wall (**6–4**). The shoe molding then covers the staples (**6–5**).

6–4 The baseboard has been installed and painted before the vinyl floor covering is laid and stapled.

6–5 The shoe molding covers the staples along the edge of the vinyl flooring.

Ceramic-tile flooring can be installed before or after the baseboard. If the baseboard is to cover the edge of the tile, the tile is usually laid first, running close to the wall. The baseboard then covers the end of the tile. The tile can be grouted before the baseboard is installed. Another technique is to install the baseboard flush with the floor and let the tile butt it. A thin line of grout is placed between the tile and baseboard (**6–6**). No shoe molding is required.

6–6 The baseboard was installed first and then the ceramic tile laid, butting against it and leaving a small grout line. No shoe molding is added.

Laminate flooring is laid loose over a soft pad. It is often referred to as a "floating floor." Usually, it is laid up to ⅜ inch from the wall before the baseboard is installed (**6–7**). The baseboard then covers the end. A shoe molding is optional.

Carpet is usually laid after the baseboard has been installed and all the painting is finished (**6–8**). The baseboard is installed ¼ to ⅜ inch above the floor so the carpet can be tucked beneath it (**6–9**). Usually, no shoe molding is used. The baseboard must be finished before the carpet is installed.

One final consideration is where to place the miter saw when cutting baseboard and other trim. If the finished floor has not been installed, it can rest on the subfloor. If a finished floor has been installed, often the saw is placed in the

6–7 This laminate floor has been installed to within ⅜ inch of the drywall. Notice the spacers at the wall. It is laid over a foam rubber pad. The baseboard will be installed after the installation is complete.

6–8 The carpet is being laid up to the baseboard. No shoe molding is required.

6–9 After the carpet is secured to the tackless strip along the wall, it is tucked under the baseboard.

garage or porch. This is very inconvenient. If you do work in the room, precautions must be made to protect the floor from the legs of the saw stand. Also, scraps must continually be swept up, since walking on them would scar the finished floor.

PAINTING THE WALLS & TRIM

Another decision that has to be made is when to paint the walls and trim. While it is best to prime the wood trim on the front and back before installation, this is often not done. It can be painted or sealed in the garage or in a room that has only the subflooring exposed. Natural and stained trim requires special consideration. See Chapter 9 for information on this type of finishing.

Some prefer to paint the interior walls before the trim is installed. In this case, the painters return and paint the trim or finish the natural or stained trim. This is a bit slower than installing the trim before painting the walls, but generally produces a neater job.

If the walls and trim are painted after the trim has been installed, the latex paint on the walls is often sprayed or rolled on, which requires that the trim be masked so it is protected. The trim is then painted with some form of enamel. Sometimes, the trim has the final finish coat applied before it is installed. After it has been installed, the nails are set and the holes filled and touched up with the paint. There may also be chips at the miter joints and areas that need caulking and have to be painted by hand.

CHECKING THE FRAMING

Check the framing of the door openings and walls to see if it is level and plumb. Also, check to see if the finish wall material, such as drywall, is flush with the door or window frames. The way to correct these and other problems is covered in Chapters 4 and 5.

BASEBOARD INSTALLATION TECHNIQUES

There are many styles of stock baseboard from which to choose. Some are shown in Chapter 1. In addition to single-piece baseboard, baseboard can be constructed from an assembly of two or more moldings (**6–10**). Others are custom-cut to the architect's profile.

The door casing should already be in place, as should cabinets that may be butted by the baseboard.

Begin by planning the sequence to be used for installing the baseboard. While carpenters use different procedures, the plan shown in **6–11** is typical. Each piece should have a square cut on one end and a cope or miter on the other. Generally, the first piece installed will be the one on the longest wall. This piece may have square cuts on each end, and will usually require a splice with a scarf joint. This is discussed later. Outside corners will be mitered and inside corners coped, as shown in **6–16**.

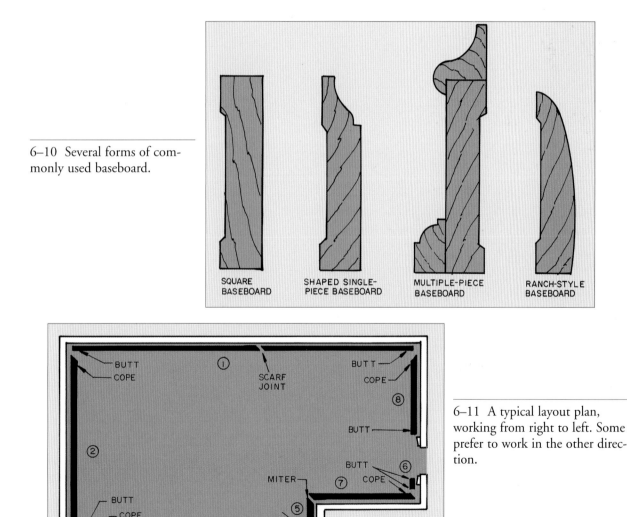

6–10 Several forms of commonly used baseboard.

6–11 A typical layout plan, working from right to left. Some prefer to work in the other direction.

Some work to their left around the room, while others prefer to work to the right. Some prefer to measure and mark the length of each piece, cut them all to length, and then begin installation; others mark, cut, and install each piece separately. When measuring to the end to be mitered or coped, mark the direction of the cut on the stock to remind you which way to set the saw (**6–12**). If the cut is to be a square cut for a butt joint, mark the location with an S (for square). If a corner is found to be not square (90 degrees), mark the size of the angle required on the back of the baseboard. Note whether it is an inside or outside corner.

In **6–11**, the decision to work to the left was made. The longest piece, No. 1, was spliced and butted to the wall. The adjoining pieces were coped on one end and given a square cut for a butt joint on the other. Another square cut was made where the baseboard butts the door casing. Notice that the outside corner was mitered.

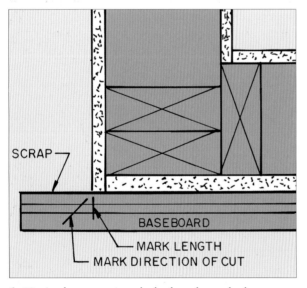

6–12 At the corner, mark the length on the baseboard and show the direction in which the miter should be cut.

MEASURING BASEBOARD LENGTH

Some carpenters prefer to cut a long baseboard to its exact length, while others will cut it 1/16 to 1/8 inch longer to get a tight fit. Shorter pieces can be cut to the exact length, or just a bit shorter. It is important to measure the length accurately to within 1/32 to 1/16 inch. Long lengths will require two people to handle the tape. The length can also be measured by placing a piece of baseboard along the wall and marking its length. Use a utility knife to get the most accurate mark (**6–13**).

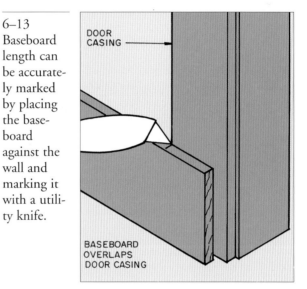

6–13 Baseboard length can be accurately marked by placing the baseboard against the wall and marking it with a utility knife.

Since the baseboard butts against them, door casings, cabinets, bookcases, fireplace surrounds, window seats, and other built-in items, they must be installed before it. Stair skirtboards are often tied into the baseboard. If there are to be heat registers mounted on the wall, hopefully they are already in place. If not, the heating contractor should mark off exactly where they will be installed. Observe if there are any changes in floor level. For example, a solid-wood floor may butt a vinyl-covered floor. Provision must be made for the difference in thickness if you want to keep the top of the baseboard level. If the difference is minor, the baseboard can be trimmed narrower where needed.

SPLICING MOLDING

When molding has to be spliced, it makes the best finished appearance if it is cut on a 45-degree angle, forming a scarf joint (**6–14**). If you use a butt joint instead, it can be filled with caulking and is fairly hidden when painted; however, over time, the molding will change in length and the joint may open. If a scarf joint opens a little, it is not as noticeable.

The joint should be located at a stud, so there is adequate nailing surface to close and hold the joint (**6–15**). One side is nailed to the stud. The other is nailed through the joint into the stud. Use two nails in each piece. It helps to lay a bead of carpenter's glue in the joint before nailing.

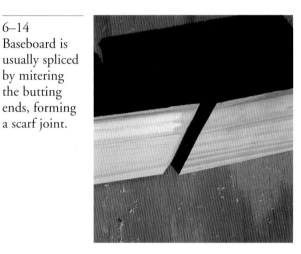

6–14
Baseboard is usually spliced by mitering the butting ends, forming a scarf joint.

MITERING AND COPING INSIDE CORNERS

Inside corners may be formed by mitering the joining moldings or by coping one of them. Coping provides a better result because as the moldings move, a crack at the corner is not apparent, while an open miter is obvious. Coped corners are used on all types of molding, such as baseboards, shoes, crowns, and chair rails.

6–15 Locate the scarf joint at a stud so the joint can be nailed into a solid backing.

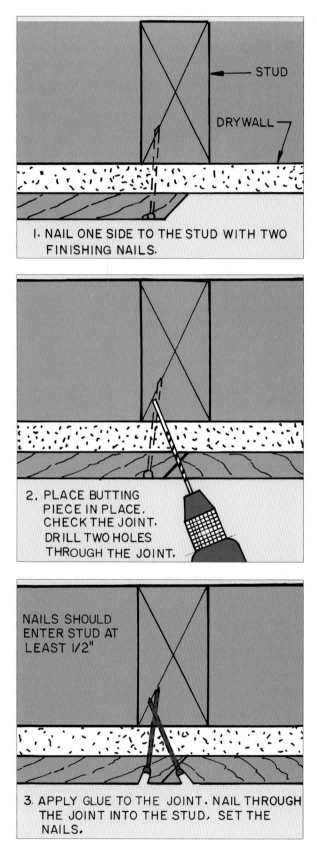

STUD

DRYWALL

1. NAIL ONE SIDE TO THE STUD WITH TWO FINISHING NAILS.

2. PLACE BUTTING PIECE IN PLACE. CHECK THE JOINT. DRILL TWO HOLES THROUGH THE JOINT.

NAILS SHOULD ENTER STUD AT LEAST 1/2"

3. APPLY GLUE TO THE JOINT. NAIL THROUGH THE JOINT INTO THE STUD. SET THE NAILS.

One of the moldings is butted to the wall. The other has the end coped, and fits over the profile of the first piece (**6–16**).

Begin by cutting a miter on the end to be coped (**6–17**). Detailed steps for cutting this miter are in Chapter 7. Cut the molding an inch or so longer than required. This provides some extra wood so the coping saw can get started cutting the profile, and provides an allowance if it becomes necessary to trim the cope. The miter should slope in the same direction as it would if an inside miter were to be cut.

Next, cut the profile using a coping saw (**6–18**). The edge formed by the miter is the line

6–17 Miter the end of the piece to be coped. Courtesy DeWalt Industrial Tool Company

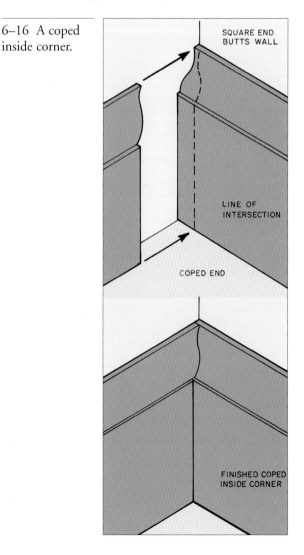

6–16 A coped inside corner.

SQUARE END BUTTS WALL

LINE OF INTERSECTION

COPED END

FINISHED COPED INSIDE CORNER

6–18 Start coping at the narrow, shaped edge of the baseboard.

of the profile to be cut. The coping saw is kept approximately **perpendicular** to the front of the molding and, as it saws, follows the profile formed when the molding was mitered.

It helps as you cut the profile to make the cuts from the back of the molding toward the profile (**6–19**). This allows the first portion cut to fall away and makes it easier to cut the next part of the profile. Be careful as you cut to keep the edge of the profile sharp and free of any chips or saw marks.

Continue cutting through to the profile and making back relief cuts until the cope cut is fin-

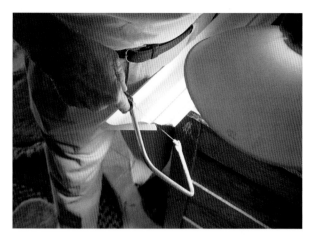

6–19 It helps to make relief cuts, removing parts of the profile as they are cut.

6–20 Smooth and adjust the curves so the cope fits the baseboard and has a clean, sharp profile.

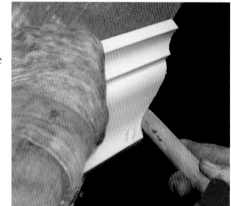

ished. Then place it against the other molding to see how it fits. It will often require a little filing to make a close fit (**6–20**). Use half-round and round files of different sizes to fit the curves being adjusted. Some prefer to angle the cope cut a little extra so the back edge, which is not seen, is slightly behind the front profile edge. This allows the front profile edge to rest firmly against the other molding. The finished, coped joint will have a sharp profile (**6–16**).

Place the coped end tightly against the adjoining baseboard. Since the coped piece is cut a little longer than needed, it can be pressed against the wall, forcing the joint closed. Nail it to the studs and bottom plate. It can be glued if you choose to do so.

MITERING OUTSIDE CORNERS

The common way to finish outside corners is with a miter joint (**6–21**). This is used for all kinds of moldings. Detailed steps for cutting miters are given in Chapter 7.

When marking the length of a piece that ends in an outside corner, prepare the end that butts the wall. Depending upon the layout, this could be a butt joint or a coped end. Then, with this piece in place and extending beyond the outside

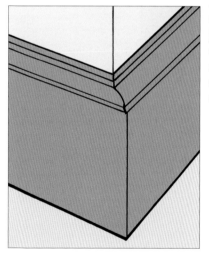

6–21 Miter joints are used on outside corners and are cut at 45-degree angles.

corner, mark the location of the inside edge of the miter with a knife or sharp lead pencil (**6–22**). The distance could also be measured with a tape. This mark is on the side of the baseboard next to the wall (**6–23**). Mark both pieces that form the corner.

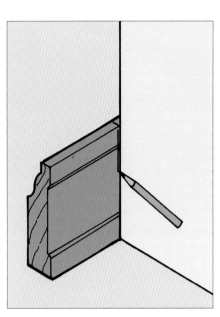

6–22 Lay the base-board along the wall and mark the length. This mark is the inside edge of the miter, and is placed against the miter-saw fence.

When cutting the miter, place the inside face of the baseboard against the fence and set the saw on the desired angle (6–24).

Since many corners are not perfectly square, some adjustments are usually necessary. Some people prefer to cut a piece of scrap stock to check the corner, or check it with a square. Slight adjustments in the angles may be necessary when cutting the finished baseboard.

After cutting the baseboard miter joint, make a trial fit (6–25). Trim it if necessary to get a closed point (6–26). Then nail one side in place.

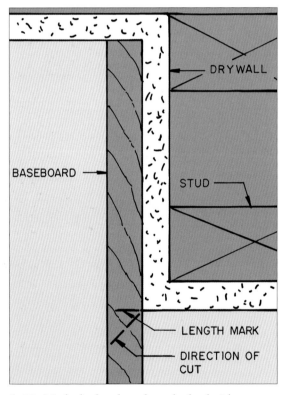

6–23 Mark the baseboard on the back side next to the wall. Mark the direction of the miter.

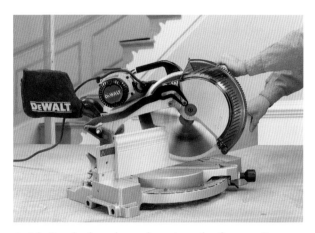

6–24 Set the length mark against the fence, adjust the saw to the proper angle, and cut the miter.
Courtesy DeWalt Industrial Tool Company

6–25 After cutting the miter joint, place it against the corner to check the fit. It may require some minor adjustment before it is nailed.

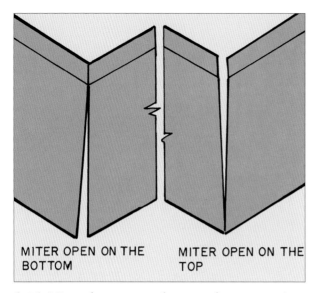

MITER OPEN ON THE BOTTOM   MITER OPEN ON THE TOP

6–26 Miters that open on the top or bottom require some trimming to get them to close.

Nail into the wall studs. You can find them by tapping on the wall or by looking for some slight dimples in the drywall where the nails were placed. Lay a bead of carpenter's glue along the miter joint and install the adjoining piece. Some prefer to lay a bead of carpenter's glue on the mitered ends and pin the joint together with a brad from a power nailer before nailing the second piece to the wall (6–27).

6–27 Glue and nail the miter closed.

INSTALLING MITERLESS BASEBOARD

There is a wood product available that eliminates the need for miters or coping. It uses separate internal and external corner pieces against which the baseboard is butted (6–28). The corner pieces are glued and nailed to the wall. The baseboard is cut to length with 90-degree ends and fitted between the corner pieces. The nails are countersunk and filled. This product can be stained, painted, or varnished.

6–28 Miterless baseboard uses corner pieces on the inside and outside corners. The baseboard is cut to length to fit between them. Courtesy Ornamental Moldings

INSTALLING SQUARE-EDGE BASEBOARD

Square-edge baseboard is installed by mitering exterior corners and either butting or mitering inside corners. If base cap and shoe moldings are used, they are mitered on exterior corners and coped on inside corners.

FASTENING THE BASEBOARD

Softwood baseboard can be installed by nailing through the finish wall material into the studs with two finishing nails long enough to give good penetration into each stud and plate. Usually 6d or 8d finishing nails are adequate.

Drive one nail, about ½ inch from the bottom, into the bottom wall plate (6–29). This will be covered with the shoe molding or carpet. The other nails will have to be set and the holes filled. Nail into each stud and near the ends of each piece. If splitting is a problem, drill small holes for the nails. The procedure for nailing wide baseboard is shown in 6–30.

Hardwood often requires that the nails are driven into predrilled holes. This helps prevent splitting and bent nails. The nails can be hand- (6–31) or power-nailed. The use of a power nailer reduces splitting and speeds up the installation.

Before starting to nail, you must locate each stud. There are a number ways this can be done. A stud-finder tool can be used. Also, you can tap on the drywall until you get a solid sound. Drive a nail at this point to see if it hits the solid stud or breaks through the drywall. If it misses, move the nail over and try again until you hit the stud. Try to get near the center of the stud. Mark it on the wall or floor. This will leave a series of nail holes in the wall, but they will be covered by the baseboard. After you locate one stud, you can measure every 16 inches and you will be very near the other studs. If there is an electrical box, it is usually mounted on a stud.

If the wall has a very slight curvature, the baseboard can usually be pulled tight to it as it is nailed. An extra nail or two may help.

Some traditional interiors have plinth blocks and rosettes on the door openings. The casing and baseboard butt against these blocks. Plinth blocks and rosettes of various designs are available from companies that manufacture moldings. While there is no hard and fast rule, the block is often wider than the casing and higher than the baseboard (6–32). It should be thicker than both so that it has a reveal on the sides where the molding butts it. There is more information on baseboards and plinths in Chapter 5.

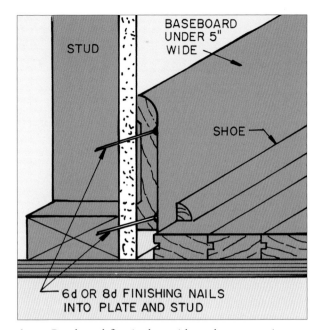

6–29 Baseboard five inches wide and narrower is nailed to the stud and plate.

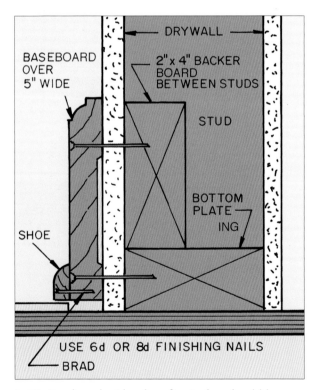

6–30 Baseboard wider than five inches should be nailed to backer boards between the studs.

6–31. Baseboard can be hand- or power-nailed.

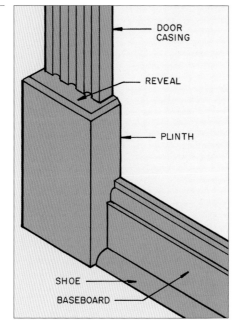

6–32 The baseboard butts the plinth and usually has a reveal.

DOOR CASING

REVEAL

PLINTH

SHOE

BASEBOARD

SECURING SHORT PIECES

Often, when the baseboard ends in a corner, one baseboard piece will need to be very short. If it is several inches long, it can be nailed with long, thin brads. However, a very small piece will split, and should be fitted into place with some carpenter's glue on the back (**6–33**). The butting baseboard piece will usually hold it even if glue is not used.

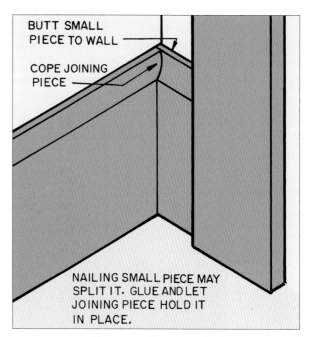

BUTT SMALL PIECE TO WALL

COPE JOINING PIECE

NAILING SMALL PIECE MAY SPLIT IT. GLUE AND LET JOINING PIECE HOLD IT IN PLACE.

6–33 Very small pieces of baseboard will often split if nailed. Glue the small piece and let the adjoining baseboard piece hold it in place.

INSTALLING SHOE MOLDING

Shoe molding[1] is installed after the finished flooring is in place. If the floor is wood and has not been sanded and finished, cut and fit the molding and lightly tack it in place. The floor finishers will remove it so they can sand closer to the baseboard. It will be permanently installed after the wood floor has been finished. Not all flooring materials require shoe molding.

The molding is measured and cut in the same way as described for baseboard. Inside corners are coped and outside corners are mitered. Most carpenters prefer to nail it to the baseboard rather than the floor. It is nailed to the baseboard with

[1] Shoe molding is a smaller, curved molding (about ¾ inch tall) that is located directly in front of the baseboard. In addition to serving aesthetic purposes, it protects baseboards from being chipped and dented by furniture, vacuum cleaners, and other things that move around at floor level.

brads as shown in **6–30**. If it is nailed to the floor and the baseboard moves a little, unfinished wood above the shoe molding will be exposed. If the molding is nailed to the baseboard, a crack can appear at the floor; however, this is not very noticeable. When nailed with power-driven brads, the shoe molding is less likely to split and the work goes much faster (**6–34**).

Where the shoe molding meets the door casing, the end should be mitered (**6–35**).

6–34 A power brad nailer will rapidly secure the shoe molding and is less likely to split it. Courtesy Paslode, an Illinois Tool Works company

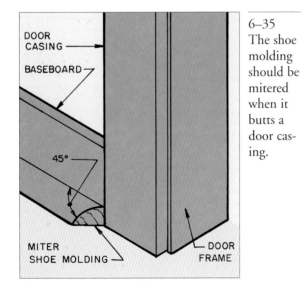

6–35 The shoe molding should be mitered when it butts a door casing.

INSTALLING BUILT-UP BASEBOARD

Multipiece baseboards are usually built around square-edge stock, upon which a small piece of molding or a cap molding is installed (**6–36**). Sometimes, a wider molding is used instead of the typical quarter-round shoe molding.

Install the square-edge baseboard using a miter joint at the outside corners and a butt joint at inside corners (**6–37**). Then, install the cap and shoe moldings by coping the inside corners and mitering the outside corners (**6–38**).

If the baseboard ends without butting another molding or surface, consider coping the cap molding. This gives a neat finished appearance (**6–39**).

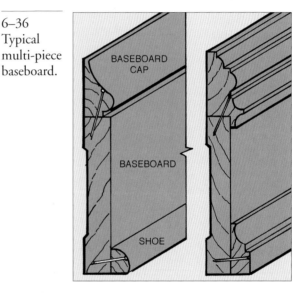

6–36 Typical multi-piece baseboard.

MAKING A MITERED MOLDING RETURN

When a molding ends without meeting another molding or surface, the exposed end grain is visible. The molding end can be cut square or mitered and carefully sanded and painted (**6–40**). While this does not present the best appearance, it does give a finished appearance. A better way of finishing an end is to make a mitered return.

Begin making the mitered return by cutting the exposed end on a 45-degree angle (**6–41**).

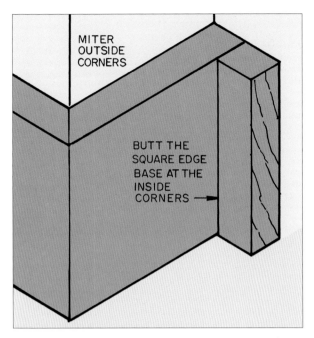

6–37 Square-edge baseboard is mitered on outside corners and butted on inside corners.

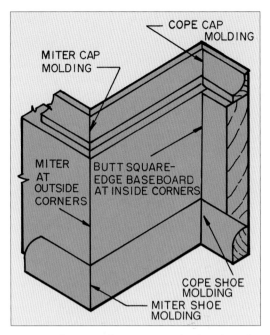

6–38 When installing square-edge baseboard, cope the cap and shoe moldings at inside corners and miter them at outside corners.

6–39 When the baseboard terminates with the end exposed, finish it and the shoe molding with a mitered return. If a base cap is used, it should be coped.

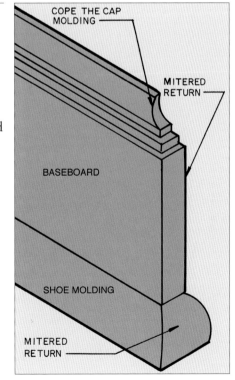

6–40 When a baseboard or other molding terminates without butting other molding or casing, its end can be mitered and carefully sanded.

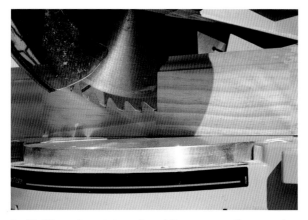

6–41 To make a mitered molding return, first cut the baseboard on a 45-degree angle.

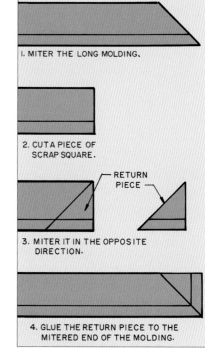

6–43 Cut the return piece from a piece of molding.

1. MITER THE LONG MOLDING.

2. CUT A PIECE OF SCRAP SQUARE.

RETURN PIECE

3. MITER IT IN THE OPPOSITE DIRECTION.

4. GLUE THE RETURN PIECE TO THE MITERED END OF THE MOLDING.

Then, install the molding to the wall by nailing it into each stud with finishing nails. Drill holes for the nails if there is danger of the molding splitting (6–42).

Next, cut the return piece from a piece of the same type of molding (6–43). Check the piece on the mitered end to be certain it is the correct size (6–44). Then coat it with carpenter's glue (6–45) and press it in place. Usually, you will need to tack it with a very thin brad or clamp it until the glue dries. It is a very fragile piece, and can break as you try to fasten it (6–46).

6–44 Check the return on the mitered end.

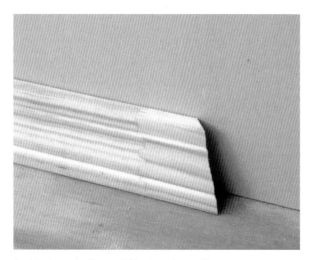

6–42 Attach the molding to the wall.

6–45 Coat the return piece with carpenter's glue and press it in place. Tack it with a very thin brad.

6–46 The finished mitered return provides a smooth, attractive end to the molding.

BUTTING A STAIR SKIRTBOARD

Frequently, the baseboard will have to butt the bottom end of the stair skirtboard[2] (6–47). Typically, the stair is installed before the finish interior trim is started, so the baseboard will have to be cut to butt the angle left on the skirtboard. This will usually be either a perpendicular butting end, or one on an angle.

6–47 Frequently the baseboard will butt the stair skirtboard as it reaches the floor.

[2]A skirtboard is a finished trim board used on either open or closed sides of the stair as a decorative accent.

INSTALLING CHAIR RAIL

Chair rail[3] is available in various stock profiles; however, if something larger or different is wanted, it can be assembled from two or more moldings. Some of the many profiles available are shown in Chapter 1.

If a stock chair-rail molding is used, it will generally be rather narrow. It is recommended that blocking be let in between the studs before the drywall is installed, to provide a solid nailing base (6–48). Narrow chair rail can bow slightly, leaving small gaps along the wall. Larger moldings, made from an assembly of moldings or a

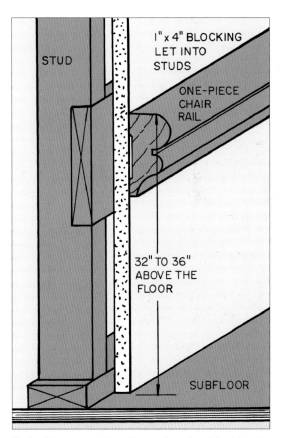

6–48 Narrow chair rail requires that blocking be let into the studs to provide an adequate nailing surface.

[3]Chair rail is a wooden molding on a wall that is installed around a room at the level of a chair back.

wide custom-made molding, may be strong enough to be nailed to the studs (**6–49**). Chair rails are also used to cap wainscoting. This is discussed in Chapter 8.

Chair rails are generally installed 32 to 36 inches above the floor, as shown in **6–48**. Begin by locating a level line around the room at the desired height. This can be done with a laser level (**6–50**) or a chalk line (**6–51**). Place single-molding chair rail on the line and tack it to the blocking every 8 to 10 inches with small finishing nails. Two nails are usually required at each location.

If a two-piece rail (as shown in **6–49**) is used, first place the piece of molding on the line and nail it. Then install the cap molding. There is often some slight waviness in the drywall. The painters should caulk this gap before finishing. Usually, the chair-rail parts are primed before they are cut and installed. Be certain to prime their backs, as well. This prevents moisture from entering the molding, reducing the chance of it warping over time.

The chair rail will usually meet a door or window casing, or a cabinet. The chair rail can simply butt a cabinet. If the chair rail is narrower than the casing, it can also butt that. If it is wider than the casing, it can be notched around it (**6–52**).

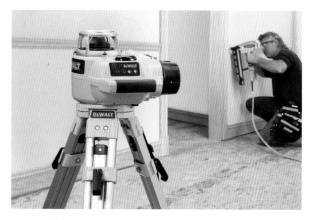

6–50 The chair rail is being located with a laser level. It is being secured with a power nailer. Courtesy DeWalt Industrial Tool Company

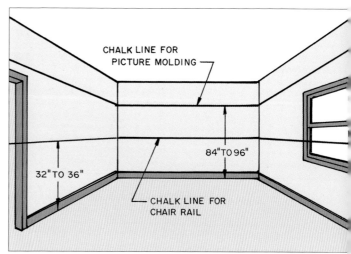

6–51 Chair-rail and picture-molding locations can be marked by snapping a chalk line.

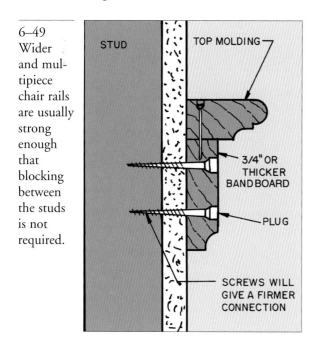

6–49 Wider and multipiece chair rails are usually strong enough that blocking between the studs is not required.

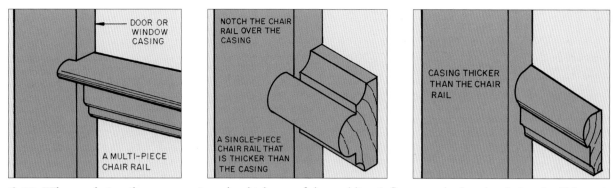

6–52 When a chair rail meets a casing, the thickness of the molding influences whether the chair rail will butt the casing, or be notched around it.

INSTALLING PICTURE MOLDING

Picture molding is installed in the same manner as chair-rail molding. It must have blocking let in between the studs, and the interior carpenters need to be aware of this before the drywall installers arrive (6–53). Picture molding is usually installed 84 to 96 inches above the floor. If the room has eight-foot ceilings, picture molding can also serve as a small crown molding.

Since picture molding is used to support heavy pictures, it needs to be securely nailed or, better still, screwed to the blocking. Screws with oval heads can be left exposed. Flat-head screws should be set below the surface, but be certain the molding chosen is thick enough to allow this to be done.

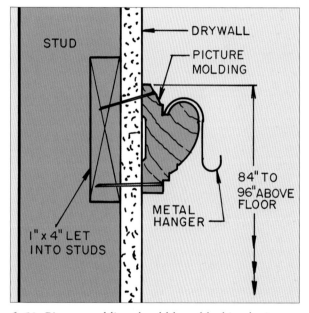

6–53 Picture molding should have blocking let into the studs so it can be securely fastened, enabling it to carry the heavy pictures.

CHAPTER 7

Crown Molding, Built-Up Cornices & Flat Ceiling Molding

rown molding and built-up cornices provide a decorative transition between the ceiling and wall and add to the ambience of the room (7–1). In classical early homes, they were of a style that reflected the architectural style of the house, such as Georgian or Victorian. Crown molding is combined with other moldings to produce detailed head cornices over doors and windows, and find use in many other applications such as fireplace surrounds (7–2) and cabinetry (7–3).

Various flat moldings are also used to decorate the area between the ceiling and the wall (7–4).

Crown molding is more difficult to install than other moldings, but the basic techniques, joints, and corners are similar. A study of the moldings in Chapter 6 will give a basic understanding that is helpful when installing crown molding.

7–1. This multiple-piece cornice was constructed using a crown molding at the ceiling.

7–2 The top of the mantel on this fireplace surround was framed with a crown molding, forming a shelf.

7–3 These wall cabinets are enhanced by topping them off with a crown molding made of the same wood.

7–4 Flat molding is also used to provide a transition between the wall and ceiling.

FLAT CEILING MOLDING

Flat ceiling molding is installed against the drywall in the same manner as baseboard. The top edge butts the ceiling. It may be a single piece of material or an assembly of two or more moldings (7–5). It is often used when the wall is covered with paneling.

Flat ceiling molding is mitered at outside corners, and coped at inside corners if they have a shaped bottom edge; otherwise, it can be butted.

The ceiling cornice in 7–4 was built of flat wood members that form a box at the corner, and is trimmed with flat molding.

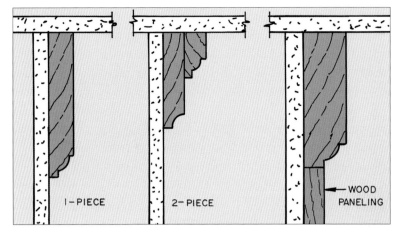

7–5 Flat molding can be used to form attractive trim at the ceiling.

CROWN MOLDING

Crown molding is made to cover or decorate a joint formed where the interior wall and ceiling meet. While crown molding is a single shaped molding, it is often combined with other moldings to form interior and exterior cornices (7–6). Often, a smaller sloped molding, called a bed molding, is used in an assembly (7–7). Compound assemblies of molding along the ceiling are also referred to as a **cornice**.

7–7 Ceiling trim can be assembled from many stock moldings, such as square stock and this bed molding.

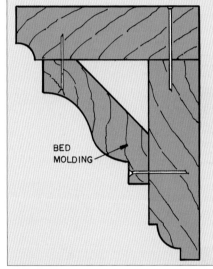

BED MOLDING

7–6 Multiple-piece crown moldings combined with other moldings are used to form large, complex cornices.

FINDING THE SPRING ANGLE

Crown moldings are made to slope away from the wall and touch the ceiling. The angle of slope is the **spring angle**. A few of the common types of crown molding are shown in Chapter 1. They most often have a spring angle of 38 or 45 degrees from the wall (7–8). Other angles are available, so consider this when you buy the molding.

If you have the molding on the jobsite, you can find the spring angle by holding the molding against the wall with the bottom edge down, as it will face when installed. Place the arm of the

Bosch Angle Finder against the wall and move the other arm until it is parallel with the face of the molding (7–9). Read the number on the LED (light-emitting diode) display on the tool and subtract 90 degrees.

Another way to find the spring angle is with a protractor, available at building-supply dealers. Place the crown molding against the wall in the normal position. Place the protractor behind it and open it until it is parallel with the molding. Read the angle directly from the scale (7–10). Do not subtract 90 degrees.

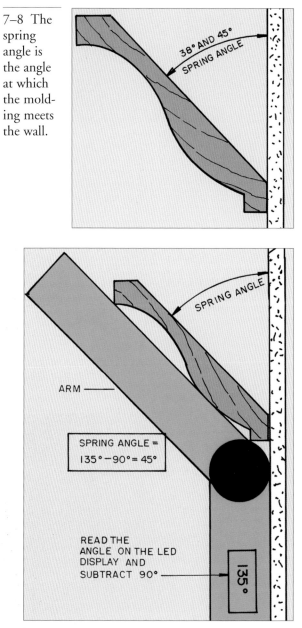

7–8 The spring angle is the angle at which the molding meets the wall.

7–9 To find the spring angle, hold a piece of the molding against the wall and place the Bosch Angle Finder against the wall with one arm along the molding. Read the angle on the LED display and subtract 90 degrees.

7–10 The spring angle can be found with a special protractor available at most building-supply dealers.

PREPARING TO INSTALL CROWN MOLDING

Crown molding up to about four inches wide may be nailed to the studs and ceiling joists (7–11). Nail it through the flats on the top and bottom edges (7–12). Wider crown moldings or wide multipiece ceiling moldings require that blocking be installed between the studs and ceiling joists before the drywall is installed (7–13). This also lets you nail the molding wherever you want, without having to locate a stud or ceiling joist. If this did not happen, you can install a shaped wood nailing strip on the surface of the drywall ceiling and nail the crown molding to it as shown in 7–14.

When installed on walls that run parallel with the ceiling joists, blocking is nailed on top of the wall top plate and extended out enough to provide a nailing surface (7–15). This blocking must be installed before the drywall is in place.

7–11 This small crown molding is being nailed into a ceiling joist. **Courtesy Paslode, an Illinois Tool Works Company**

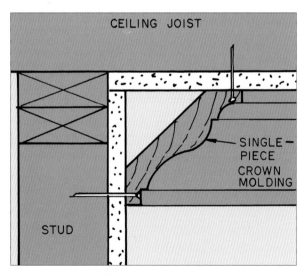

7–12 Small, single-piece crown moldings can be nailed to the studs and ceiling joists.

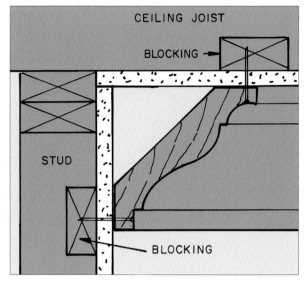

7–13 Wide moldings can be nailed to blocking between the studs and ceiling joists. This must be installed before the drywall.

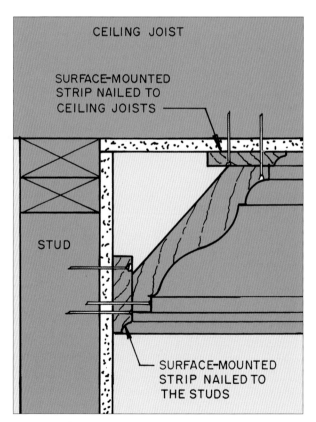

7–14 Surface-mounted wood strips can be used for nailing crown molding.

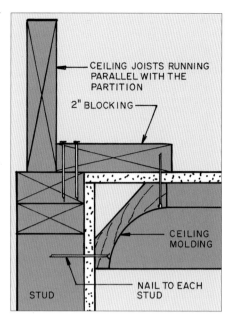

7–15 Blocking to support crown molding can be installed on the top plate and extend out along walls that run parallel with the ceiling joists.

CEILING JOISTS RUNNING PARALLEL WITH THE PARTITION

2" BLOCKING

CEILING MOLDING

STUD

NAIL TO EACH STUD

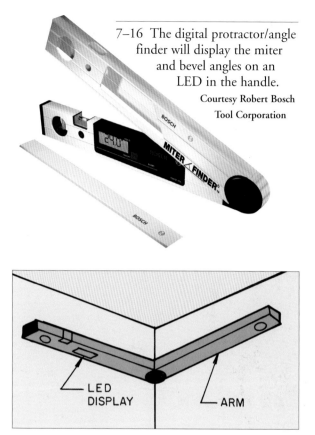

7–16 The digital protractor/angle finder will display the miter and bevel angles on an LED in the handle.
Courtesy Robert Bosch Tool Corporation

LED DISPLAY

ARM

7–17 The angle of every corner of the room should be checked before the crown molding is cut. The Bosch Angle Finder is one tool used for finding the existing angle.

FINDING CORNER, BEVEL, & MITER ANGLES

Crown molding is mitered at outside corners and mitered or coped at inside corners. Coped corners produce the best results over the years. Since most walls do not meet at exactly a right angle, the miter frequently has to be adjusted.

Establish the **angle of the corner**. This can be measured with a tool like the Bosch Angle Finder (**7–16**). Follow the instructions of the manufacturer. Basically, this involves setting the spring angle on the LED (light-emitting diode) display, placing the Angle Finder in the corner, and opening the arms flat against the wall. Press the designated buttons, and the **bevel** and **miter angles** will appear on the display. This system is designed to cut crown molding when it is placed flat on the table of a compound miter saw (**7–17**).

Miter- and bevel-angle charts are available that give information you need to cut 45- and 38-degree spring angles. Compound-saw manufacturers can supply angle charts for a wide range of wall angles, typically from 67 to 179 degrees. Data for the 38- and 45-degree crown molding at a 90-degree corner is given in **Table 7–1**.

The **miter angle** is the angle as measured from the bottom edge of the crown. The **bevel angle** is the angle as measured from the back face of the crown. The miter is the angle the blade makes with the fence. The bevel is the degrees the blade is tilted.

CUTTING CROWN MOLDING WITH A STANDARD MITER SAW

To make a compound-miter cut with a standard miter saw, you must place the molding on the saw table with its bottom edge up against the fence and its face side out. The top edge must rest flat on the saw table (**7–18**), so that it is actually upside down and backwards to the way it will be

TABLE 7–1

Settings for Compound Miter and Bevel Angles for a 90-Degree Angle for a Compound Miter Saw with the Moldings Flat on Table

Molding Spring Angle	Cope on the Right End of the Molding With the Top Edge Against the Fence		Cope on the Left End of the Molding With the Bottom Edge Against the Fence	
	Miter Angle	Bevel Angle	Miter Angle	Bevel Angle
45/45°	35.26°	30°	35.26°	30°
52/38°	31.62°	33.86°	31.62°	33.86°

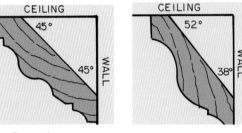

Courtesy Delta International Machinery Corporation

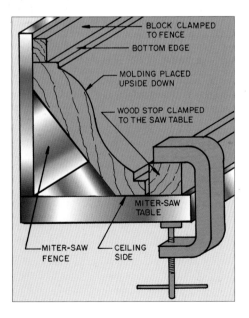

affixed to the wall. This will let you make compound-miter cuts using the settings on the saw scale to the left or right of the center mark.

Detailed instructions for mitering an **outside corner** are given in 7–19 and 7–20. The **right-side miter** is shown in 7–19 and the left-side miter in 7–20.

Detailed instructions for mitering **inside corners** to be coped are in 7–21 and 7–22. Instructions for cutting the **left-side cope miter** are given in 7–21, and the **right side cope miter** in 7–22.

7–18 Position the crown molding with its bottom edge up when cutting it. Some clamp a block to the miter-saw table or fence to prevent the molding from slipping.

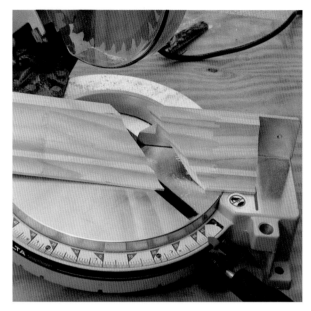

7–19 **Cutting the crown molding for the right side of the miter for an outside corner.** Place the molding with the bottom edge up against the fence. Set the blade to the 45-degree mark on the right and the molding to the left. The scrap end is on the right.

7–20 **Cutting the crown molding for the left side of the miter for an outside corner.** Place the molding with the bottom edge up against the fence. Set the blade to the 45-degree mark on the left and the molding to the right. The scrap end is on the left.

7–21 **Mitering the crown molding for the left side of the cope for an inside corner.** Place the molding with its bottom edge up against the fence. Set the blade to the 45-degree mark on the right. Place the molding to the right of the blade. The scrap end is to the left.

7–22 **Mitering the crown molding for the right side of the cope for an inside corner.** Place the molding with its bottom edge up against the fence. Set the blade to the 45-degree mark on the left. Place the molding to the right of the blade. The scrap end is to the right.

CUTTING CROWN MOLDING WITH A COMPOUND MITER SAW

The compound-miter saw makes it much easier to cut miters and bevels, because the molding can be placed flat on the table and the saw can be set to the required miter and bevel angles. On some cuts, the top edge of the molding is placed next to the fence. On others, the bottom edge is next to the fence.

Detailed instructions for mitering **outside corners** are given in **7–23** and **7–24**. The **right-side miter** is described in **7–23**, and the **left-side miter** in **7–24**.

Detailed instructions for mitering **inside corners** to be coped are given in **7–25** and **7–26**. The **left-side cope miter** is described in **7–25**, and the **right-side cope miter** in **7–26**. **The steps to cope this miter are in Chapter 6.**

SPLICING CROWN MOLDING

Long walls will require that two or more pieces of molding be spliced. The pieces are joined with a scarf joint (7–27) which, when carefully cut and secured, is relatively unobtrusive. On paint-grade molding, any minor openings can be caulked and sanded. Scarf joints on stain-grade molding can have cracks filled with a wood filler of a color similar to that of the finished molding. A thin line will show, but this is not noticeable (7–28). The ends forming the joint are mitered as described for mitering outside corners.

Many people prefer to make the splice before the pieces are put up on the wall. Be certain the total length of the spliced piece is longer than required. This allows some material to be removed if the splice needs some recutting, and provides extra material when cutting the ends on each wall.

Other people install one piece on the wall with the mitered end sloping into the wall. The adjoining piece is slipped into this, and they are glued and nailed to the wall studs and ceiling joists. This can work for narrow molding, but is not the best technique for moldings over three inches wide.

The scarf joint is secured in various ways, depending upon the preferences of the finish carpenter. One good way to preassemble the joint is to glue and staple a thin wood backing over the joint. A typical backing is either ¼-inch plywood or a thin, solid-wood strip. The thickness can be greater if there is room behind the molding so it clears the wall and ceiling (7–29). If a rosette (corner blocking) is used for the installation, a very thin backing strip is necessary.

Some people prefer to glue and staple the backing strip to one piece, apply glue to the joining surfaces, close the joint, and glue and nail the backing strip to the second piece. If you do this, staple the backing strip into the thicker part of the molding. Other people will drive a few brads through the face of the molding into the backing strip. If you do this, set the molding on a flat surface until the glue has dried. Remember, the joint is a weak spot in the molding, so it must be handled carefully or it may open up. Usually three or more people are required to handle room-length spliced crown moldings.

Another technique for securing a scarf joint is to machine grooves in the joint surfaces to receive plates. The plate is glued into the machined groove on one piece (7–30). The adjoining piece is covered with glue and joined to the first. A backing strip is stapled over the joint, as described earlier. Plate joiners are discussed in Chapter 3.

MAKING A PLAN

In order to plan for the materials needed and the end joints on each piece, make a plan showing the order of installation and the cuts to be used on each end. While finish carpenters have various approaches, the following example is typical (7–31). The first piece installed is usually the longest. The ends are butted to the walls, and a scarf joint is often required. Many carpenters like to work from left to right and proceed with

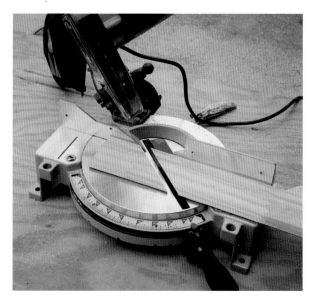

7–23 Cutting the crown molding for the right side of an outside corner. Place the molding with its back side flat on the table and its top edge against the fence. Set the blade 45 degrees to the right and tilted to the specified bevel angle. Place the molding to the right of the blade. The scrap end is to the left.

7–24 Cutting the crown molding for the left side of an outside corner. Place the molding with its back side flat on the table with its bottom edge next to the fence. Set the blade 45 degrees to the left and tilted to the specified bevel angle. Place the molding to the right of the blade. The scrap end is to the left.

7–25 Mitering the crown molding for the left side of the cope for an inside corner. Place the molding with its back side flat on the table and its top edge against the fence. Set the blade 45 degrees to the right and tilted to a 45-degree bevel. Place the molding to the left of the blade. The scrap end is to the right.

7–26 Mitering the crown molding for the right side of the cope for an inside corner. Place the molding with its back side flat on the table and its bottom edge against the fence. Set the blade 45 degrees to the left and tilted to a 45-degree bevel. Place the molding to the left of the blade. The scrap end is to the right.

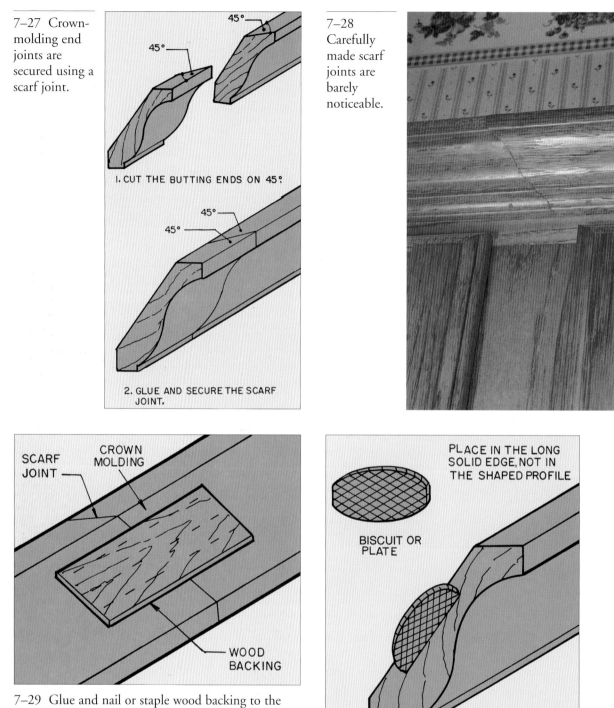

7–27 Crown-molding end joints are secured using a scarf joint.

1. CUT THE BUTTING ENDS ON 45°.

45°
45°

45°
45°

2. GLUE AND SECURE THE SCARF JOINT.

7–28 Carefully made scarf joints are barely noticeable.

SCARF JOINT

CROWN MOLDING

WOOD BACKING

7–29 Glue and nail or staple wood backing to the rear of the crown molding. Also glue the scarf joint.

PLACE IN THE LONG SOLID EDGE, NOT IN THE SHAPED PROFILE

BISCUIT OR PLATE

7–30 Scarf joints can be joined with plates. Glue the joint surface and glue the plate in the groove.

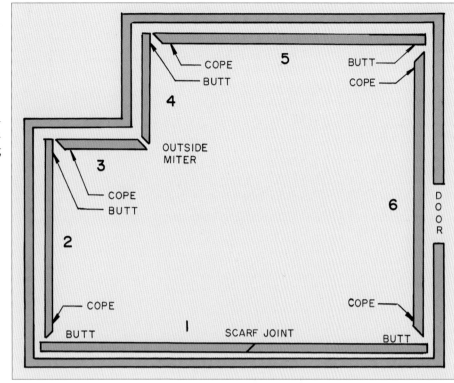

7–31 One way to lay out a plan for the crown molding in a room.

the installation in this direction. Notice how on most pieces one end is coped and the other is butted against the wall. The last piece has to be coped on both ends.

You can also write the lengths required for each piece on your plan. This way, you will be able to refer to it when cutting each piece.

FINDING CROWN MOLDING RISE

Before the layout line is run around the wall, it is necessary to find the rise of the crown molding. The rise is the vertical distance from the top of the molding to the bottom when it has been installed. This can be found by placing a small piece of molding against a square and reading the rise in inches as shown in **7–32**.

INSTALLING CROWN MOLDING

Locate a line on the wall around the room that is down from the ceiling a distance equal to the rise of the molding. This can be done by snapping a

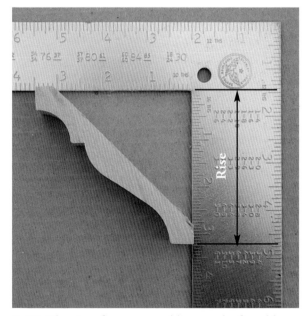

7–32 The rise of a crown molding can be found by placing it on a square in the installed position and reading the vertical distance on the scale.

chalk line (7–33) or by measuring down every few feet to a mark made using a wood block cut to the length of the rise (7–34). The bottom edge of the molding is placed on this line. This keeps it level. Since ceilings tend to have bows, this will have to be taken into consideration as the rise line is located. The rise line can also be located with a laser level.

Small crown molding can be installed by nailing it to each ceiling joist and wall stud (7–12). While it can be hand-nailed with finishing nails, a power nailer is faster and less likely to split the molding (7–35). This requires that each joist and stud be located and marked. Possibly, the easier way to do this is with a stud finder. There are a number of different types on the market. Another alternative is to install blocking, as shown in 7–13 and 7–14.

A more secure way to install the molding is to use corner blocks. Small moldings can use short corner blocks (7–36). Some people install short lengths of blocking, spaced every 12 to 16 inches, while others run a continuous strip the entire length of the wall (7–37). The blocking is nailed to the top plate.

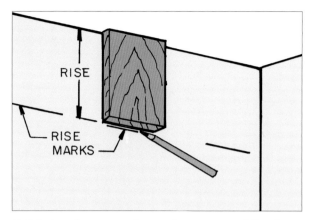

7–34 Another way to locate the crown molding is to make a measuring block the length of the rise and mark a series of dashes along the wall.

7–35 The best way to nail crown molding is with a power nailer. Courtesy Paslode, an Illinois Tool Works company.

To install the molding, nail it into the blocking (7–38). This way, you do not have to nail it to the studs or joists. Wide crown moldings can use a two-piece blocking arrangement (7–39).

If a scarf joint is being used, be certain to leave a gap for the reinforcing strip on the back.

HANDLING BOWED CEILINGS & WALLS

As you place the crown molding against the ceiling and wall and line it up with the rise line, you will frequently notice that the ceiling and wall

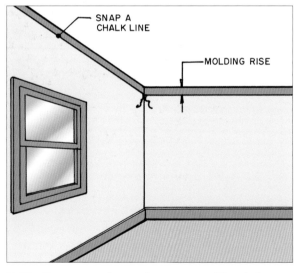

7–33 One way to locate the crown molding below the ceiling is to measure down the rise and snap a chalk line.

INTERIOR TRIM: MAKING, INSTALLING & FINISHING

7–36 A good way to install crown molding is to nail corner blocks to the stud and top plate. Then nail the molding to the corner blocks.

7–37 Instead of nailing small individual corner blocks, some prefer to nail longer strips or a continuous corner strip. This lets you nail the crown molding wherever it seems best, rather than nailing only at the studs.

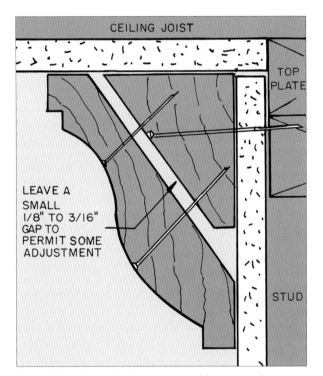

7–38 Nail through the crown molding into the blocking. Leave a small gap between the blocking and molding to allow space for necessary adjustments.

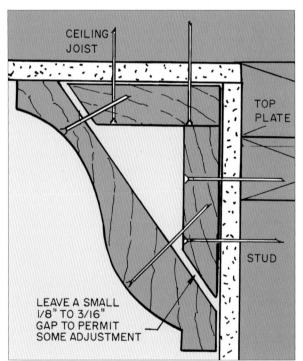

7–39 Wide crown molding can be mounted with surface-mounted, two-piece blocking.

are not perfectly flat. They have an occasional bow that will make a gap between the molding and ceiling or wall and force it out of level. To check the ceiling, measure up a fixed distance from the floor in each corner, say, eight feet (7–40). Run a chalk line between these points. Mark any low spots. To locate the bottom edge of the crown molding, measure down from the low spot a distance equal to the molding rise. Run the rise line through this point.

If the gap between the ceiling or wall and paint-grade molding is small, it can be caulked. If the gap is too large to caulk or occurs with stain-grade molding, have the drywall contractor feather a layer of drywall compound out over the ceiling to reduce the size of the gap (7–41).

Since the crown molding must be perfectly straight, small gaps along the wall or the ceiling should be shimmed with thin strips of wood (7–42). This will hold the edge straight and flat as it is nailed. If it is not shimmed, the pressure of the nailing will bow the molding. Cut the shim flush with the bottom edge of the crown molding. Caulk the gap between the molding and the wall or ceiling.

INSTALLING MITERLESS CROWN MOLDING

Miterless crown moldings do not need inside and outside miters because they are supplied with internal and external corner blocks. The crown molding is cut to length with square ends, and butts these corner pieces as shown in 7–43. The corners and moldings are glued and nailed to the wall.

INSTALLING POLYURETHANE CROWN MOLDING

Polyurethane crown molding is available in a wide range of sizes and profiles and produces large, complex moldings (7–44) that, if made from wood, would require the assembly of many individual pieces of molding. Polyurethane molding is supplied as a single piece that is ready to install. Some of these moldings are so wide they cannot be cut with a standard miter saw. They must be cut and installed following the recommendations of the manufacturer. Some must be cut face up on the table of a miter saw. Others must be cut in a vertical position, held against the saw fence.

Following are instructions for installing polyurethane crown molding supplied by one manufacturer: Place the molding in the room at least 24 hours before it is installed. Avoid storing or installing it in extreme temperatures (below 40 and over 85 degrees Fahrenheit). The surface upon which it is to be installed should be clean, dry, and free from oil, grease, or other impurities. Remove old wallpaper and repair loose, damaged plaster. Porous surfaces should be primed, and glossy surfaces sanded to improve bonding.

The installation steps include:
1. Lightly sand the surfaces of the molding that will be painted. Wipe the surfaces clean with a cloth dampened with mineral spirits (7–45). This will help the paint adhere.
2. Accurately measure the length of the walls that will have molding, and cut the molding 12 to 18 inches longer if there will be an outside miter. This gives some extra material for cutting the miter.
3. Start installation in the most conspicuous outside corner and end in the least conspicuous inside corner. Since often the pattern will not exactly match at the last corner, this will be the least noticeable spot.
4. Miter the molding as instructed by the manufacturer (7–46).
5. Use only the glue and other materials supplied by the manufacturer (7–47). Apply a ⅛-inch bead of the glue to the molding surfaces that will touch the ceiling and wall (7–48).
6. Cut the molding about ⅛ inch longer than required for every five feet of length. Press the piece against the wall and ceiling, and nail or

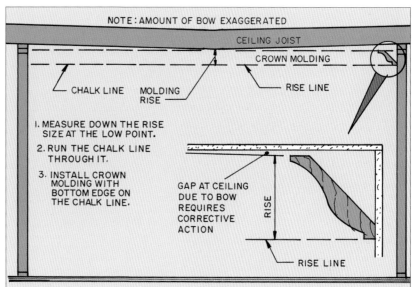

7–40 If the ceiling has a slight bow, locate the low spot. Measure down the molding rise and run a chalk line through it. Install the bottom of the molding on this line. Take corrective action to fill the gap at the ceiling.

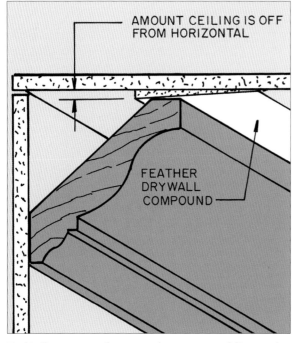

7–41 Larger gaps between the crown molding and the drywall ceiling can be concealed by feathering out a layer of drywall compound.

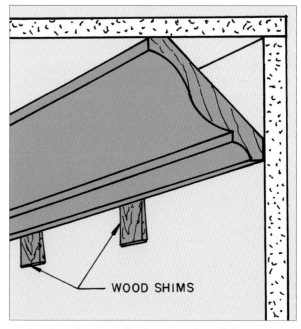

7–42 Gaps between the bottom edge of the crown molding and the drywall are secured with shims and then caulked. Tap in the shims until the molding is straight. Cut them flush with the bottom of the molding. Caulk the gaps.

7–43 Miterless crown molding uses interior and exterior rosettes. The molding is cut to fit between them.
Courtesy Ornamental Moldings

7–44 Polyurethane crown moldings provide complex moldings in a single strip. This makes installation fast and easy. Courtesy Architectural Products by Outwater, LLC (800-835-4400)

7–45 Lightly sand the surfaces of the molding to be painted.
Courtesy Architectural Products by Outwater, LLC

7–46 Miter the molding following the manufacturer's instructions.
Courtesy Architectural Products by Outwater, LLC

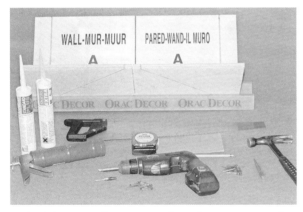

7–47 The manufacturer supplies a kit containing the adhesive, filler, tools, and instructions. Courtesy Architectural Products by Outwater, LLC

7–48 Lay a ⅛-inch bead of glue along edges that will touch the wall.
Courtesy Architectural Products by Outwater, LLC

screw each end to the wall stud and ceiling joist (**7–49**). Then, press the piece against the wall in the center. Nail or screw it in the center. Countersink the fasteners (**7–50**).

7. Make the splices. Do this by cutting the butting pieces square, forming a butt joint. Again, cut each piece ⅛ inch longer than required. Install one section and apply adhesive to the end (**7–51**). Then install the other piece. Since it is a bit long, you will need to press the pieces against the wall, forcing the joint together. Nail the joint to the studs and ceiling joist. Allow the adhesive to cure several hours, and remove any excess with a razor blade or sharp chisel.

8. Miter the inside and outside corners. The joint is bonded with adhesive as described for splices.

9. After the molding has been installed and the adhesive cured, fill the nail and screw holes, the spaces in the joints, and any dented or damaged areas with the manufacturer-supplied filler (**7–52**). This can also be used to fill any deviations between the molding, ceiling, and wall. Wipe excessive filler off with a wet cloth. When the areas are dry, lightly sand them.

Polyurethane moldings can be finished with oil- or latex-based paint. Application with a brush is recommended. Do not spray latex-based paint.

7–49 Press the crown molding against the wall and nail or screw it to the stud and ceiling joist. Courtesy Architectural Products by Outwater, LLC

7–50 Set the nails or countersink the screws. The holes will be filled later. Courtesy Architectural Products by Outwater, LLC

7–51 To splice the molding, cut the end square, apply the glue, and press the pieces together. Nail the joint securely so it is tight.
Courtesy Architectural Products by Outwater, LLC

7–52 After the adhesive has cured, fill any nail and screw holes, cracks, and damaged areas with the supplied filler. Courtesy Architectural Products by Outwater, LLC

CROWN MOLDING, BUILT-UP CORNICES & FLAT CEILING MOLDING

WALL SHELVES

There are many unusual but interesting applications for the use of crown molding. A little original thinking will produce some very interesting results. One idea is to use the molding to make small wall shelves, as shown in **7–53**. The molding is notched, to hold the top shelf and the back. The bottom board is glued to the bottom of the crown molding (**7–54**).

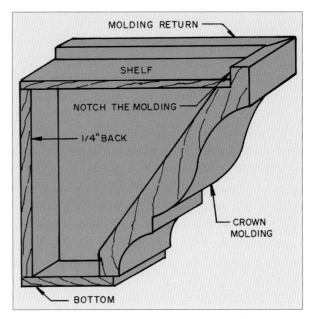

7–54 The crown molding forming the wall shelf is notched to receive the thin shelf board and back.

7–53 This crown molding has mitered ends with a return to the wall. It provides a most interesting, small wall shelf.

CHAPTER 8

Wainscoting & Molding Wall Panels

For hundreds of years, walls covered partway up with wood have been popular. This type of trim is referred to as **wainscoting** (**8–1**). Wood wainscoting produces a very strong wall. Wood has historically been the easiest material to find to cover inside walls. The finished walls are also very attractive, especially as shaped, raised panels have became popular.

8–1 Solid-wood, raised-panel wainscoting has been used for hundreds of years. Notice how it blends in with the cabinets. **Courtesy Ornamental Moldings**

In the early days of the United States, pine and oak were the most commonly used woods for wainscoting. In more recent times, almost every species available has been used. The species, the way it is finished, and the design of the wall panels provide the major architectural element of the room. Early classical styles of homes, such as those built in New England, tended to develop a particular style that is often copied on restoration jobs and new classic style homes of that period.

In many cases, the wall above the wainscoting is also paneled with wood (**8–2**).

Today, wainscoting takes many forms and designs, and is used in houses of all architectural styles. In addition to solid wood (**8–1**), veneered plywood, painted medium-density fiberboard, fabric-covered panels (**8–3**) of oriented strandboard or plywood, polyurethane panels (**8–4**), and vinyl covering over embossed paper panels are also used. The designs of the panels can be very decorative, such as those used in Early American Colonial homes, or very simple, to fit into a contemporary house. A typical example is the use of vertical boards, such as tongue-and-groove paneling or other types of flat wood panels. Usually, these have shaped edges that provide a shaped vertical seam. These can be stained and finished with a clear coating,

8–2 The wall above the raised-panel wainscoting is paneled with wood. Notice that the raised panel is a different species of wood than the frame.

8–3 This wainscoting has a wide chair rail over a fabric-covered wall panel. Courtesy Mr. and Mrs. James Sazama.

8–5 Wood wainscoting is often used to trim the wall along a stair.

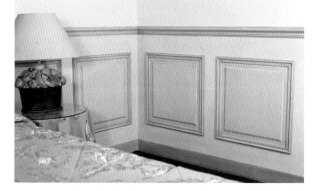

8–4 The polyurethane raised panels are bonded to the wall, producing the appearance of wood raised-panel wainscoting. Courtesy Architectural Products by Outwater, LLC (800-835-4400)

such as lacquer or polyurethane, or painted to match the other interior trim.

Wainscoting is also used on the walls along a staircase (**8–5**). In addition to being decorative, it also resists damage better than drywall.

Wainscoting is typically 36 inches high, but in some period houses it goes up to about two-thirds of the ceiling height (**8–6**). Remember, a room with stained wainscoting or paneling that covers most of the walls will make the room appear smaller and darker. Consider the overall appearance as decisions are made. Sometimes a chair rail will be a better solution. A natural wood or a light stain will reduce the sometimes overpowering influence of the wainscoting. Paint-grade wainscoting materials can be painted to match the trim, and possibly the color on the wall above it. A light-colored paint will reduce the dominance of the wainscoting, yet permit the design and shadows it produces to provide a very attractive wall (**8–7**).

SOME TRADITIONAL WAINSCOTING

Early American colonial design for panel wainscoting was brought to the colonies from Europe. It was a good way to finish interior walls because wood was the major material available. Since wood is also a good insulator, this was a major benefit.

8–6 While 36-inch-high wainscoting is most often used, higher panels suit the design of some rooms. The wainscoting in the room upon which this illustration was based was painted.

STANDARD WAINSCOTING TYPICALLY 32" TO 36"

HIGH WAINSCOTING TYPICALLY 60" TO 72"

8–7 Painted, solid-wood wainscoting below a plaster wall. It is about 100 years old.

The early settlers tended to consider excessive decoration an unnecessary extravagance, so they used paneling with a very simple design (8–8). As the colonies grew, people from a number of countries moved here, bringing other designs and cultures, which resulted in more decorative designs.

Wainscoting designs were greatly influenced by the Victorian era, which occurred during the reign of Queen Victoria of England (1840–1900). Rapid design changes during this time lead to the use of elaborate, deeply sculpted moldings, and heavy ornamentation. A typical decorative wainscoting is shown in 8–9.

During the 1800's, a very conservative wainscoting using beadboard evolved. It was frequently used in such rooms as kitchens, pantries, and bathrooms, where decoration was not so important as filling a utilitarian purpose (8–10). Today, this type of wainscoting is used to give a country farmhouse appearance to a kitchen or bath.

In the 1900's, simpler wainscoting designs became popular that reflected the style of houses being built. They featured flat panels and straight-line frames (8–11). This period in architecture is often referred to as the Mission or Modern period.

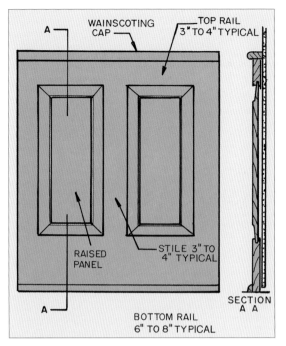

8–8 Early American colonial homes used a simple wainscoting similar to those used in Europe.

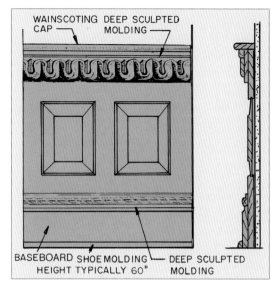

8–9 Heavily decorated wainscoting using a carved chair rail and sculpted moldings is typical of designs popular in the Victoria era.

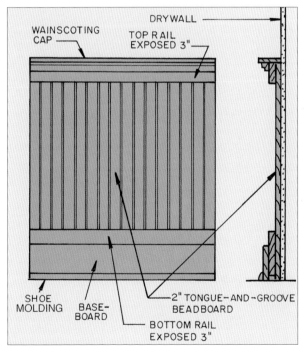

8–10 Beadboard wainscoting is used in kitchens and bathrooms whose design is inspired by American houses built in the 1800's. It is usually painted to match the trim.

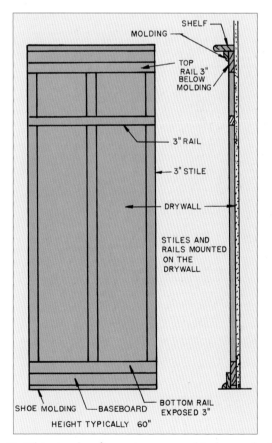

8–11 This simple wainscoting was used in houses built in the early 1900's when simpler designs became popular. It is often called Mission or Modern wainscoting.

WAINSCOTING & MOLDING WALL PANELS

BUILDING CODES

While most types of wainscoting can be secured directly to the wall studs and blocking, most building codes require that a ½-inch sheet of gypsum drywall be installed, and the wainscoting put over it. This is because gypsum drywall provides protection against fire.

ELECTRICAL & HEATING CONSIDERATIONS

As you plan the layout of any type of wainscoting, remember to consider the location of electric and telephone outlets, and heat registers. If installing paneled or tongue-and-groove materials, this is not a major problem. However, if paneled wainscoting is to be used, size the panels so the electric outlet is located within the panel, on the bottom rail, or centered on a stile (**8–12**). Since the outlet plate is generally three inches wide, the stile or rail must be wide enough to accommodate it.

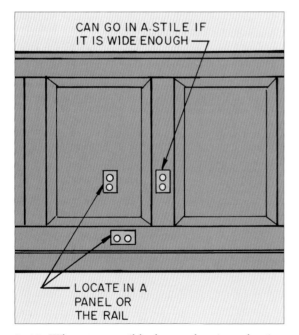

8–12 Whenever possible, locate electric outlets in a panel; however, if need be, they can be placed in the bottom rail or centered on a wide stile.

If the outlet hits a stile and cannot be moved, it will be necessary to first build a small frame from molding around the electrical box that is slightly larger than the box cover plate. Next, cut the wainscoting to fit around the frame and butt it tightly. Then add an electric box extender to bring the duplex outlet up to the surface of the wood frame (**8–13**).

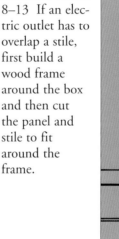
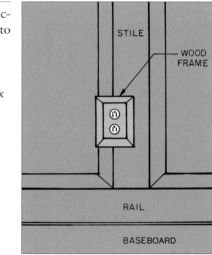

8–13 If an electric outlet has to overlap a stile, first build a wood frame around the box and then cut the panel and stile to fit around the frame.

If the wainscoting uses a flat panel and a thin overlaying stile, and the outlet is located in the stile, the stile can be notched to fit around the outlet cover (**8–14**).

If this is new construction, you might give the electricians a layout drawing of the wainscoting so they can situate the outlets within a panel. If this is not possible, or it is old construction, consider having the electrician come in and move the outlet. Since the drywall will be covered by the wainscoting, the damage to it caused by moving the outlet will not show. Be certain to repair the damage and seal all openings so the wall meets fire-resistance codes.

Since raised- or flat-panel wainscoting adds to the thickness of the wall, an outlet box extension is added to bring the duplex outlet plugs up to the surface of the panel (**8–15**). Electrical codes

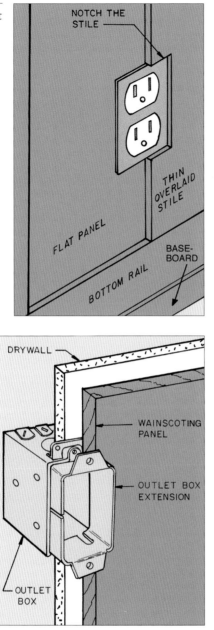

8–14 When flat panels are used with thin overlaid stiles, the stiles can be notched to fit around the outlet.

8–15 An outlet box extension is used to bring the duplex outlet flush with the surface of the wainscoting paneling.

require the front edge of the outlet box to be flush with the surface of the panel.

In new houses, the heat registers are usually in the floor. In older homes, they may be mounted up on the wall or at floor level, and provision must be made to work the wainscoting around them. Flat panels and tongue-and-groove wainscoting can be cut around them; however, if

using paneled wainscoting, you will have to cut into the bottom rail (**8–16**). If the rail is narrow, special framing will have to be made to get the panel to fit over the register.

8–16 In older houses, the heat registers may be in the wall or come through the baseboard at the floor.

CHARACTERISTICS OF FRAME & PANEL WAINSCOTING

Frame-and-panel wainscoting consists of a solid-wood frame into which some types of panel are installed (**8–17**). This paneling system is more expensive than the others in common use, but provides a depth and pattern that echoes the beauty of high-quality classical houses built in years past. The depth and shadows created produce a most attractive wall finish.

THE FRAME

The frame is solid wood and consists of horizontal top and bottom rails and vertical stiles. The rails run the full length of the room. The assembled wainscoting is screwed to the wall framing. The frame is grooved to receive the panel (**8–17**). The groove is cut ⅛ inch deeper than needed to receive the panel. This allows for expansion and contraction. The stiles are secured to the rails with tenons, dowels, or plates (**8–18**). The end stile butts the wall, or is placed against a door or window casing, as shown in **8–6**. These joints are glued, forming a rigid frame.

THE PANELS

There are a number of different types of panel typically used. Possibly the most attractive is the raised panel. This is similar to the type of construction used in homes in early America and Europe. It is the most expensive. The raised panel is a solid-wood unit, machined around the edges to produce a taper with the center standing out in relief (**8–17**).

The raised panel can be made using a shaper or powerful router mounted on a router table. Special panel-raising router bits are used to make the cuts. They are large and dangerous, so always run the stock along a fence and wear eye and ear protection. Cut slowly so you do not overload or stall the shaper or router.

The raised panel can also be cut with a table saw. Since the panel is usually quite large, the normal saw fence is too low to hold the panel perpendicular to the table. You can screw a 12-inch piece of plywood to the fence to provide support, or clamp a block to the panel as shown in **8–20**.

To cut the panel, first cut the reveal by placing the panel face down on the saw table and cutting grooves for the reveal. Since the reveal is sloped (**8–19**), the saw blade must be set to the desired angle. The reveal is usually 1/8 inch deep. Next, cut the tapered surface. Set the blade to the

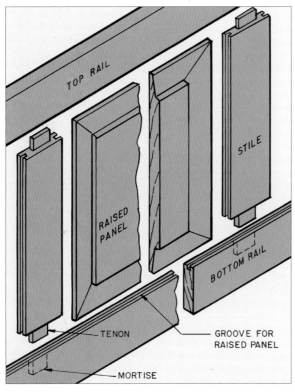

8–17 Typical construction of raised-panel wainscoting.

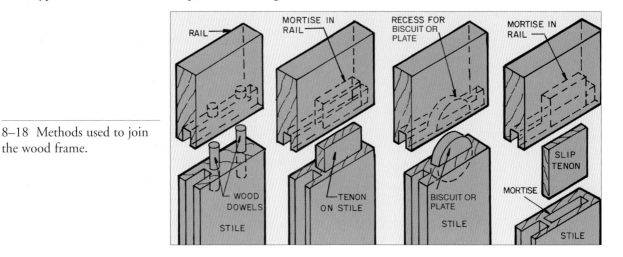

8–18 Methods used to join the wood frame.

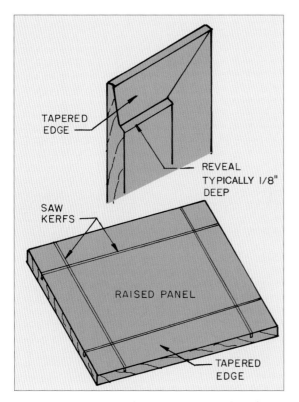

8–19 Cut the reveal, forming the shoulder of the raised panel.

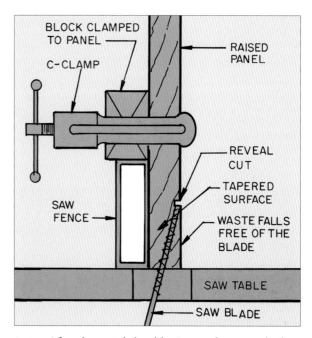

8–20 After the reveal shoulder is cut, the tapered edge of the panel can be cut on a table saw. Clamp a block to the panel. The block slides along the top of the saw fence, keeping the panel perpendicular to the saw table.

desired angle and position the fence away from the saw, as required. Always make the cut so the scrap piece falls to the outside (**8–20**). If the scrap falls next to the fence, the blade will kick it back, causing possible personal injury. The large amount of blade exposed while making the cut will often cause burned spots, so the surface will not be in a finished condition after it is cut. It will require considerable sanding. Because of this, the use of a shaper or router will produce the best results.

Frame-and-panel wainscoting is often custom-made by a local woodworking shop or a panel manufacturer. Since room sizes vary so much, it is difficult to work with stock panels. One stock raised panel is available in various sizes made from polyurethane (**8–21**). These are bonded to the drywall with a manufacturer-supplied adhesive. To make them work out, you

8–21 These polyurethane wainscoting panels are available in a range of sizes. **Courtesy Architectural Products by Outwater LLC (800-835-4400).**

WAINSCOTING & MOLDING WALL PANELS

have to adjust the spaces between the panels so they are all the same width.

The wood raised panel is inserted into the grooves in the rails and stiles. While the rails and stiles are glued together, the panels are not glued into the groove. This allows for expansion and contraction. A detail of this is given in **8–22**. An alternate technique used is shown in **8–23**.

Other types of panel commonly used include flat, overlaid, overhung, and recessed panels. Various **flat panels** can include a veneer-covered plywood or other panel product rigid enough to span between the sides of the frame without eventually bowing. It can be set in a groove or rabbet, as shown in **8–24**, for raised panels. The **overlaid panel** consists of a sturdy flat panel that has another panel bonded to it, leaving the back panel exposed around the edges. It mimics the

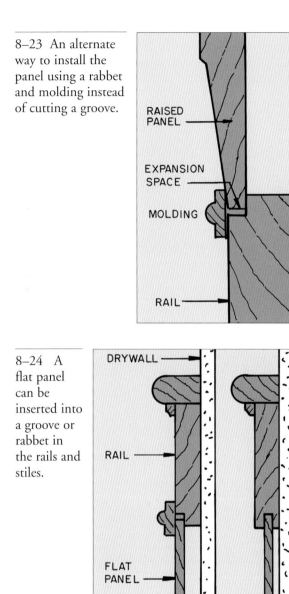

8–23 An alternate way to install the panel using a rabbet and molding instead of cutting a groove.

8–22 Typical construction of solid-wood, raised-panel wainscoting.

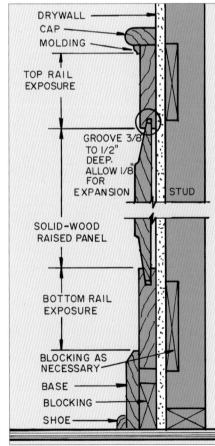

8–24 A flat panel can be inserted into a groove or rabbet in the rails and stiles.

true solid-wood raised panel (**8–25**). You can also simply mount a flat wood panel with shaped edges directly on the drywall (**8–26**).

Overhung panel construction is shown in **8–27**. These panels are typically solid wood. Notice that the exposed edge can be rounded, beveled, or cove-cut. The overhang creates an interesting shadow line along the wall. The **recessed panel** (**8–28**) also provides a raised panel effect, yet is easier to machine.

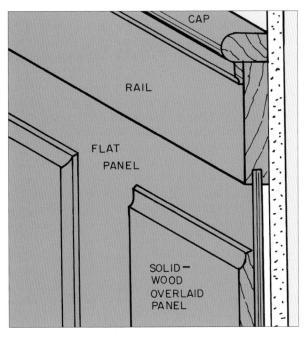

8–25 Overlay panels can be mounted on flat panels that have a wood veneer surface.

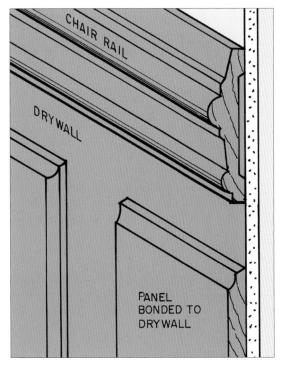

8–26 Simple wainscoting can be created by installing a chair rail and bonding overlaid panels to the drywall.

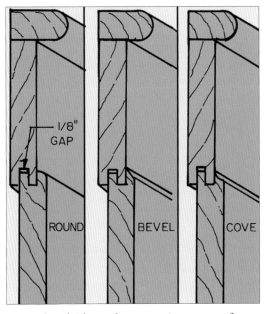

8–27 Overlaid panels are put in on top of a wood flat panel that has been installed in the grooves in the rails and stiles.

8–28 Recessed panels look much like raised panels, but are easier to machine.

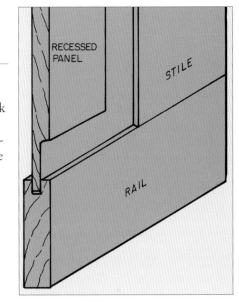

LAYING OUT THE WAINSCOTING

Begin by making some size decisions. For example, a four-inch top rail and a four- to six-inch bottom rail are commonly used. Then select the width of the panel you would like to use. This will change some as the layout is developed. Finally, select the height, which is commonly 36 inches. Now the exact layout of the panels can be made.

It helps to make a sketch of the wall (**8–29**). Measure its exact length and record it on the drawing. In **8–29**, the wall is 120 inches long. Since there is always one more stile than panel, subtract the width of one end stile from the room width. Remember, the end stiles are the selected width plus the thickness of the butting wainscoting. In **8–29**, this is the 3-inch stile plus the ¾-inch thickness of the butting wainscoting, for a width of 3¾ inches. Plan the width of the butting wainscoting stile so a 3-inch stile is exposed on both panels in the corner.

In the example in **8–29**, this left 115½ inches to be divided into stiles and panels. If a 14-inch panel and a 3-inch stile are used, each panel/stile unit equals 17 inches. Divide 115½ (the available distance) by 17 and you get 6.8 such panel units. Round the 6. 8 to 7 panel units. In order to get all the units the same size, divide 115½ inches by 7 units, which gives 16½ inches per unit. Subtract the 3-inch stile from this and the panel is 13½ inches wide.

ASSEMBLING FRAME & PANEL WAINSCOTING

After the rails, stiles, and panels have been cut, make a trial assembly without glue. This is the time to trim and fit the parts as necessary to get tight joints. If this is paint-grade wainscoting, it is recommended that the panels be primed on both sides, and the rails and stiles primed at least on the back side. If it is stain-grade work, apply one or two coats of the finish that will be used on the face side. This will seal the wood so moisture will not penetrate over the years, causing bowing and warping.

Once everything fits and is sealed, assemble the wainscoting. Let the glue dry for the recom-

8–29 Use a sketch to make some trial layouts and reach a solution as to the number of panel units and their width.

```
1. 120" – 3.75" – .75" = 115.5"
2. 115.5" ÷ 17 = 6.8 PANEL UNITS — USE 7 PANELS
3. 115.5" ÷ 7 = 16.5" PER PANEL UNIT
4. 16.5" – 3" STILE = 13.5" PANEL WIDTH
```

WALL TO WALL 120"

ONE UNIT 16.5

TOP RAIL

13.5" 3" 13.5" 3" 13.5" 3" 13.5" 3" 13.5" 3" 13.5" 3" 13.5" 3"

BOTTOM RAIL

3.75"   115.5"   .75"

mended length of time. Now you are ready to install the assembly on the wall. This produces a long, rather heavy assembly.

JOINING CORNERS

Decide how the corners will be joined. Inside corners can be butted, or secured with a groove and a tongue. Outside corners are usually mitered. They can be butted, but the edge grain of one side will be visible (**8–30**).

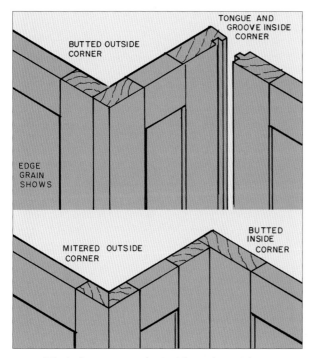

8–30 Typical ways to make inside and outside corners.

INSTALLING FRAME & PANEL WAINSCOTING

Run a chalk line or use a laser level to locate the top edge of the wainscoting. Locate it so there is a small gap at the floor (**8–31**).

When possible, first install the longest section that runs along a wall with no windows or doors. This piece is trimmed on each end so it clears

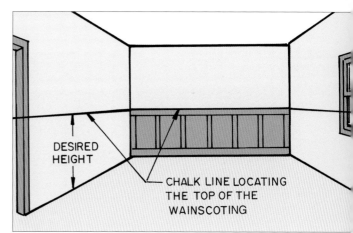

8–31 Locate the top edge of the wainscoting with a chalk line or a laser level.

the ends of the wall a little. The gaps will be covered by the butting panel (**8–30**). Line up the top of the piece with the chalk line. If the bottom rail runs to the floor, the wainscoting can be leveled by tapping wedges below it (**8–32**). If the bottom rail is placed above the floor, wood blocking is used above the floor to support the base (as shown in **8–22**). To get the wainscoting level, these blocks can be adjusted, by tapping wedges under them (**8–33**).

Secure the wainscoting to the studs and bottom plate. Paint-grade wainscoting can be secured with nails, which are set and caulked. Stain-grade wainscoting can be nailed in the same way; however, try to place the nails so they are covered by the baseboard and the wainscoting cap and molding (**8–34**). If any nails are still visible, they can be set and filled with the available filler that comes close to matching the final color.

For a very secure installation, the wainscoting can be fastened with wood screws. If exposed, they are counterbored and covered with a wood plug made of the same wood (**8–35**). If the grain in the plug is laid parallel with the grain in the rails, it is less noticeable. A possible design variation is to use wood plugs of a totally different color wood.

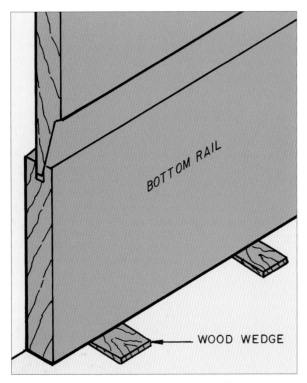

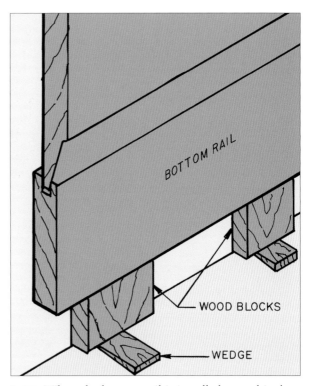

8–32 When the bottom rail runs to the floor, you can use wood wedges to level the wainscoting and hold it in position as it is nailed to the studs and wall plate.

8–33 When the bottom rail is installed several inches above the floor, it can be supported by wood blocks. These blocks can be adjusted with wood wedges to bring the wainscoting level and hold it as it is nailed to the studs and wall plate.

8–34 Typical fastening recommendations when the wainscoting cap and the base cover the nails or screws.

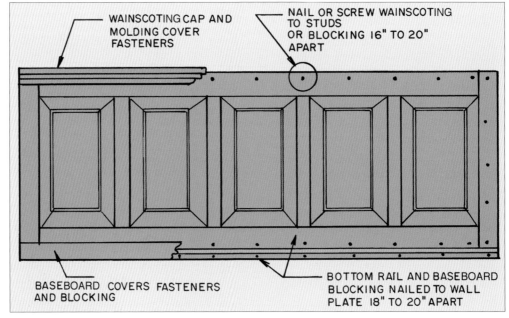

INTERIOR TRIM: MAKING, INSTALLING & FINISHING

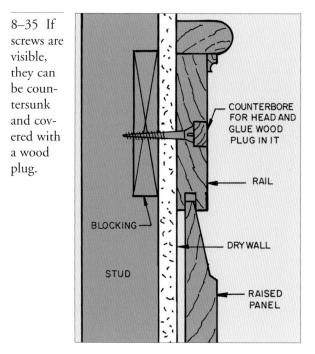

8–35 If screws are visible, they can be countersunk and covered with a wood plug.

COUNTERBORE FOR HEAD AND GLUE WOOD PLUG IN IT

RAIL

BLOCKING

DRY WALL

STUD

RAISED PANEL

INSTALLING WOOD TONGUE & GROOVE WAINSCOTING

There is quite a variety of solid-wood paneling materials available. Those most frequently used have a tongue-and-groove edge joint (**8–36**).

Walls must have blocking to which vertically applied paneling can be nailed (**8–37**). Both 1 × 4 and 2 × 4-inch blocking is used. If the paneling is applied horizontally, it is nailed through the drywall into the studs (**8–38**).

Before you start installing the paneling, make a plan. Some prefer to start at outside corners and work toward the inside corners. One reason to do this is that the outside corner is very visi-

Some prefer to install 2 × 4 blocking between the studs, as shown in **8–22**. This lets more fasteners be used and provides additional support. The assembled wainscoting sections are very heavy, so this extra security is worth considering.

Secure each butting section in the same way, working carefully to get a tight corner joint. It is a good plan to make these sections slightly longer than the wall, so that the stile at the corner can be trimmed to compensate for any slope in the butting wall. When the wainscoting butts a door or window casing, the stile must be planed to make a tight fit.

Finally, there will be one wall left. The wainscoting will have to fit tightly against it on each end. This requires careful measuring and trimming, until the assembled section slips into place. If a small gap is left on one end, you can force out the butting wainscoting with a thin wedge, closing the joint. The small gap at the wall can be covered with the wainscoting cap.

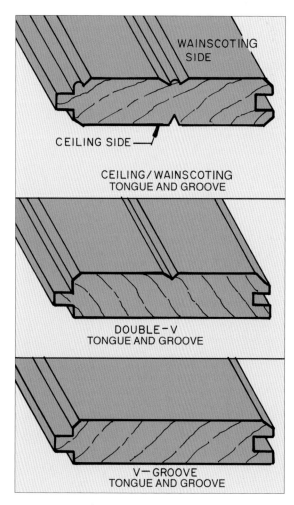

WAINSCOTING SIDE

CEILING SIDE

CEILING/WAINSCOTING TONGUE AND GROOVE

DOUBLE−V TONGUE AND GROOVE

V−GROOVE TONGUE AND GROOVE

8–36 Some of the commonly used solid-wood tongue-and-groove paneling.

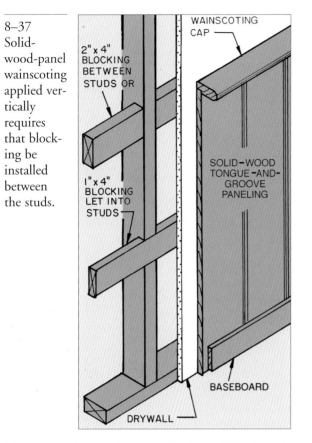

8–37 Solid-wood-panel wainscoting applied vertically requires that blocking be installed between the studs.

Labels in figure: WAINSCOTING CAP; 2" x 4" BLOCKING BETWEEN STUDS OR; 1" x 4" BLOCKING LET INTO STUDS; SOLID-WOOD TONGUE-AND-GROOVE PANELING; BASEBOARD; DRYWALL

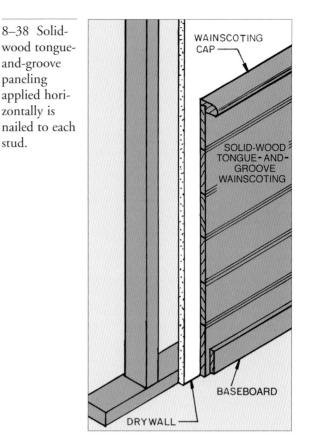

8–38 Solid-wood tongue-and-groove paneling applied horizontally is nailed to each stud.

Labels in figure: WAINSCOTING CAP; SOLID-WOOD TONGUE-AND-GROOVE WAINSCOTING; BASEBOARD; DRYWALL

ble, so beginning here with full-width boards ensures a good appearance. If a partial panel is needed to finish the wall, it is not as noticeable if it is at an inside corner (**8–39**). If there are no outside corners, begin at an inside corner and work toward doors and windows first, and then other inside corners.

Begin the installation by snapping a chalk line or using a laser level to locate the top edge of the paneling. Usually this line is the edge of the actual paneling, and does not allow for the wainscoting cap. Allow for a small gap at the floor (**8–40**).

Now check the first wall for plumb. If it is out of plumb, measure away from the wall a distance equal to the width of one board and draw a plumb line. Measure the distance from the wall to the line at the top and bottom. Mark these distances on the first piece of paneling and trim it to the line (**8–41**). Install this piece. Check to verify it is plumb.

Next, install the pieces across the wall. As you near the other wall, measure every now and then to see if the top and bottom distances are the same. If they are close to the same, you know the last panel will butt the end wall and not need a big taper (**8–42**). If the difference is large, the end panel will have to be tapered to butt the wall (**8–43**).

Tongue-and-groove paneling is blind-nailed through the tongue. The joint is not pulled tight, but left with a ¹⁄₁₆-inch space to allow for expansion (**8–44**).

Inside corners are often butted. This requires careful work to get the butting panel on the same slope as the wall. A wood molding may be used to cover this joint (**8–45**). Outside corners may be mitered or butted (**8–46**). Trim off the tongue or groove so a solid edge is available.

When the paneling meets a door or window casing, it is carefully cut and trimmed to give a

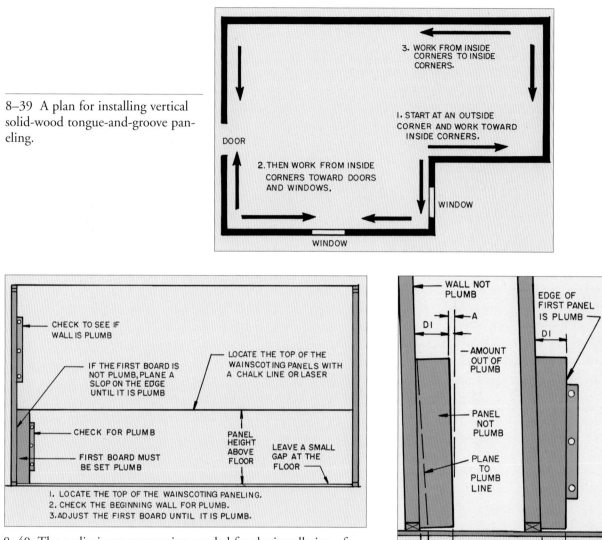

8–39 A plan for installing vertical solid-wood tongue-and-groove paneling.

8–40 The preliminary preparation needed for the installation of vertical tongue-and-groove wood paneling.

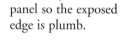

8–41 Adjust the first panel so the exposed edge is plumb.

8–42 Start installing the panels, checking to be certain they stay plumb. As you near the other wall, measure to see if the top and bottom distances are the same. If not, adjust the edge of the last panel.

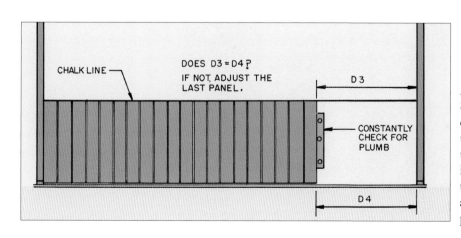

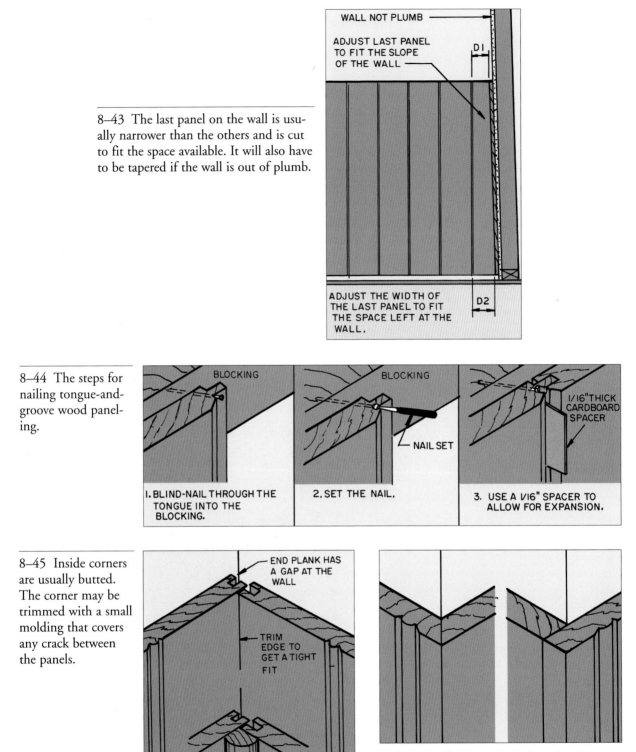

8–43 The last panel on the wall is usually narrower than the others and is cut to fit the space available. It will also have to be tapered if the wall is out of plumb.

WALL NOT PLUMB

ADJUST LAST PANEL TO FIT THE SLOPE OF THE WALL

D1

ADJUST THE WIDTH OF THE LAST PANEL TO FIT THE SPACE LEFT AT THE WALL.

D2

8–44 The steps for nailing tongue-and-groove wood paneling.

BLOCKING

BLOCKING

1/16" THICK CARDBOARD SPACER

NAIL SET

1. BLIND-NAIL THROUGH THE TONGUE INTO THE BLOCKING.

2. SET THE NAIL.

3. USE A 1/16" SPACER TO ALLOW FOR EXPANSION.

8–45 Inside corners are usually butted. The corner may be trimmed with a small molding that covers any crack between the panels.

END PLANK HAS A GAP AT THE WALL

TRIM EDGE TO GET A TIGHT FIT

SMALL MOLDING

8–46 Outside corners may be mitered or butted.

INTERIOR TRIM: MAKING, INSTALLING & FINISHING

tight fit (**8–47**). In this situation, the casing should be thicker than the paneling, so there is a reveal.

Various types of wainscoting cap molding can top off the panels, as shown in **8–48**.

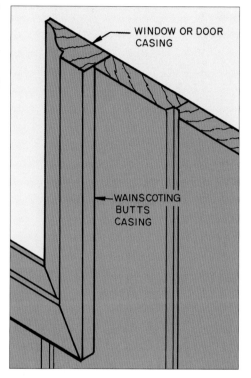

8–47 The wainscoting will butt door and window casings.

WINDOW OR DOOR CASING

WAINSCOTING BUTTS CASING

INSTALLING PLYWOOD PANELS

Plywood panels with a quality veneer on the surface are used for wainscoting construction. If the studs are spaced 16 inches OC (on center), the panels should be ¼ inch thick. If they are spaced up to 24 inches OC, use ⅜-inch-thick plywood. Panels ¼ inch thick or thinner must usually be applied over a fire-resistant backing such as gypsum wallboard (**8–49**).

Some prefer to bond the paneling to the wallboard with adhesive; however, most prefer to nail them to the studs. The nails can be set and covered with a colored wax pencil. Some manufacturers supply nails with heads colored to match the color of the paneling. The nails are installed along the top of the panel, which was located by snapping a chalk line as shown in **8–40**.

Before installing the flat panels, check to see if the end walls are perpendicular. Generally, they are off a little, so the end panels will have to be tapered to butt against the walls, as shown for wood paneling in **8–41** and **8–43**.

While there are a number of ways the inside and outside corners can be finished, probably the easiest is to install a molding on an inside

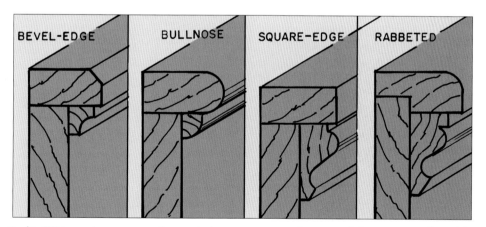

BEVEL-EDGE   BULLNOSE   SQUARE-EDGE   RABBETED

8–48 Wainscoting caps can be made in many ways using any of the many stock moldings available.

corner and a wood corner strip on an outside corner (**8–50**). When the panel meets the door and window casings, cut and plane it carefully to get a tight fit as shown in **8–47**.

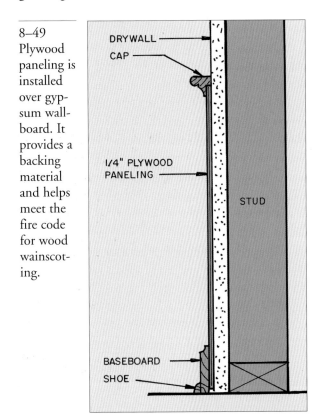

8–49 Plywood paneling is installed over gypsum wallboard. It provides a backing material and helps meet the fire code for wood wainscoting.

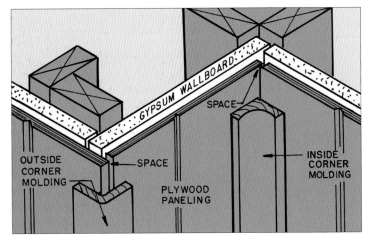

8–50 One way to finish the corners when plywood panels are used for the field of the wainscoting.

INSTALLING MOLDING WALL PANELS

Installing wall panels formed with molding is like installing a picture frame flat against the wall (**8–51**). They produce an appearance much like that of the raised wood frame-and-panel construction, but are easier to install and cost much less. A wide variety of moldings are available (**8–52**), and some manufacturers can also provide shaped corners that add a bit to the overall appearance.

8–51 The polyurethane molding on these wall panels has been painted a different color than that used on the walls. Notice the special corner moldings and the panel on the ceiling.
Courtesy Architectural Products by Outwater, LLC (800-835-4400).

PRELIMINARY PLANNING

The first thing to be decided when using wall panels for wainscoting is the height of the chair rail. While there is no standard height, 32-, 36-, and 60-inch heights are commonly used in rooms with ceiling heights of 8 to 10 feet. Avoid a height that is half the floor-to-ceiling height because then neither half of the wall will be dominant.

The height and length of the wall influence the size of the molding panel. If the panels appear too large and out of proportion to the size of the wall, windows, and doors, some of the space

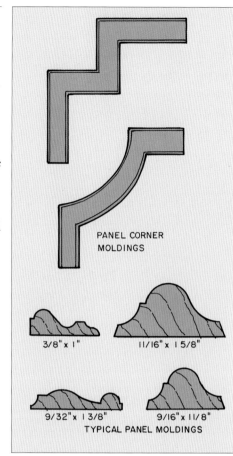

8–52 Corner moldings enhance the overall appearance of the molding wall panel. The actual moldings used vary in size and design.

PANEL CORNER MOLDINGS

3/8" x 1" 11/16" x 1 5/8"

9/32" x 1 3/8" 9/16" x 1 1/8"

TYPICAL PANEL MOLDINGS

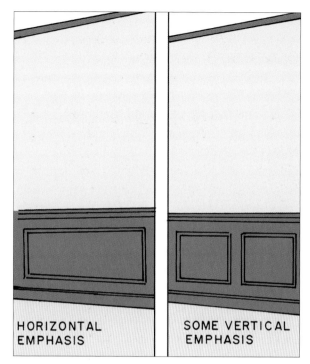

HORIZONTAL EMPHASIS   SOME VERTICAL EMPHASIS

8–53 While low wainscoting with molding panels generally uses long, rectangular panels, narrower panels will give some emphasis to the vertical appearance.

can be accommodated by adding a molding below the chair rail, enabling a smaller molding panel to be built. Also, the shape of the panel is important to the final appearance. Low chair rails work well with long horizontal panels; however, some prefer to use small panels that give some vertical emphasis (**8–53**). A high wainscoting will typically have tall, narrow panels that emphasize the height of the room. A single panel this wide would give a more massive appearance and be out of proportion with the wall (**8–54**).

SIZING THE PANELS

For hundreds of years, the most pleasing rectangular shape in architectural design has been considered to be one in which one side is 1½ times larger than the other. This is referred to as the Golden Rectangle. While you do not have to

8–54 Wainscoting in the range of 60 inches high usually uses tall, narrow molding wall panels. A single, large panel would not be suitable for many applications.

adhere to this to the very inch, it does provide a way to create a pleasing rectangular panel. Obviously, tall panels will not be able to follow this rule, but the way they are used on the wall can establish a pattern and sense of rhythm.

To get a pleasing wall, some prefer to adjust the width of the wall panels. For example, a wall that requires three panels can be enhanced by making the center panel wider (**8–55**). If a wall has two or four panels, they are usually all the same width. Long walls with five or seven panels can be enhanced by making alternate panels narrower.

Very high wainscoting requires tall, narrow panels, as shown in **8–56**. If a large single panel were used, it would become very dominant and influence the character of the room.

A wall with doors and windows requires some planning to get a balanced array of wall panels (**8–57**).

If the height of the wainscoting requires a size of wall panel that appears too tall, you can use smaller panels and run a small molding between the top of the panel and the chair rail, as shown in **8–58**. The space between the molding and panel is usually the same width as that used between panels.

Another design possibility is to install wall panels above the chair rail that are the same width as that on the wainscoting (**8–59**). This provides some interesting decorating possibilities, as shown in **8–51**. Again, consider their size and proportion so that a pleasing balance is achieved.

ESTABLISHING THE HEIGHT OF THE MOLDING PANEL

The height of the total wainscoting is a personal choice. Once it is decided, the size of the molding panel can be established. This will depend upon the width of the chair rail and baseboard selected. Typical examples are shown in **8–60**.

The length of the panel can be found much the same way as described for the raised frame-and-panel wainscoting. Measure the length of the wall. Decide on the space to be left between the panels. Now select a panel length that approaches the proportions of the Golden Rectangle. In the example in **8–60**, the height of the panel was found to be 22 inches. Therefore, the length would approach 33 inches. In actual

EQUAL-WIDTH PANELS ARE ACCEPTABLE, BUT NOT ESPECIALLY ATTRACTIVE.

VARIED-WIDTH PANELS ADD TO THE ATTRACTIVENESS OF THE WALL.

8–55 If a wall has an uneven number of panels, varying their width can improve its appearance, yet keep it in balance.

INTERIOR TRIM: MAKING, INSTALLING & FINISHING

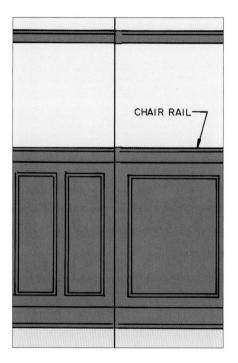

8–56 Typical molding panel arrangements for tall wainscoting. The large, single panel tends to dominate the room. The tall, narrow panels give emphasis to the vertical.

8–57 Plan the molding panel widths to work around windows and doors.

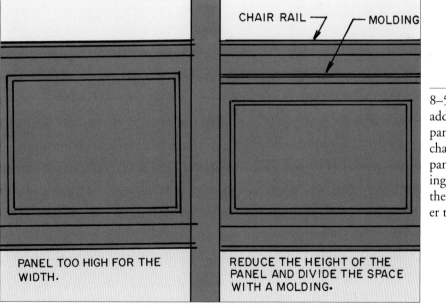

PANEL TOO HIGH FOR THE WIDTH.

REDUCE THE HEIGHT OF THE PANEL AND DIVIDE THE SPACE WITH A MOLDING.

8–58 Molding can be added between the top panel molding and the chair rail to enable the panel to maintain pleasing proportions when the wainscoting is higher than usual.

8–59 Large molding wall panels installed above the wainscoting provide a dominant architectural feature. They provide the opportunity for some interesting decorating schemes.

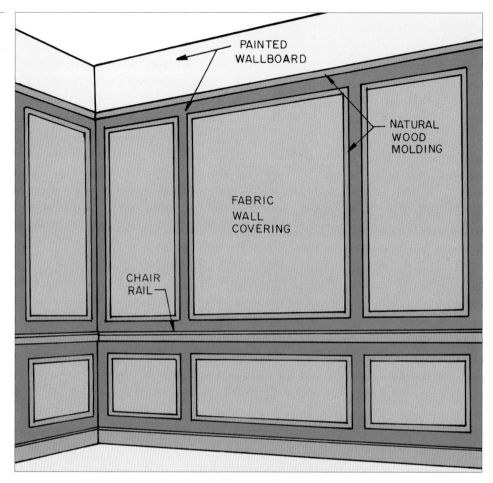

PAINTED WALLBOARD

NATURAL WOOD MOLDING

FABRIC WALL COVERING

CHAIR RAIL

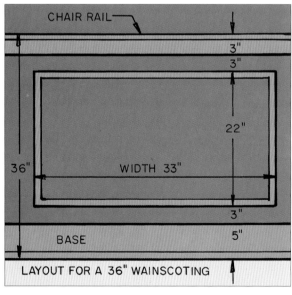

CHAIR RAIL

3"
3"
22"
36"
WIDTH 33"
3"
BASE   5"

LAYOUT FOR A 36" WAINSCOTING

8–60. The size of the molding panel depends upon the width of the chair rail and baseboard selected.

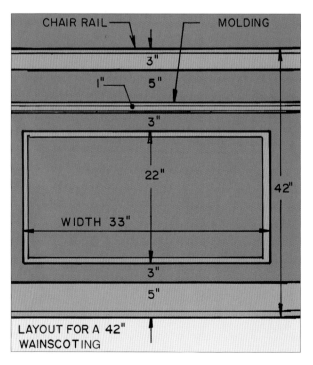

CHAIR RAIL   MOLDING

3"
1"   5"
3"
22"
42"
WIDTH 33"
3"
5"

LAYOUT FOR A 42" WAINSCOTING

practice, you will have to vary this somewhat to get the width of the panels to work out along the wall. Next, select the amount of space between the panels, the chair rail, and baseboard.

Make a sketch of the wall and record the length very accurately. Illus. **8–61** shows a layout in which there is one more open space than panel; in this case, the width of one space, 3 inches, is subtracted from the length (144 inches). The result, 141 inches, is divided into panels and spaces. Add the proposed panel length, 33 inches, and one open space, 3 inches, to get the space needed for one panel and space, 36 inches. Divide the total wall length, 141 inches, by one panel and space (36 inches) to find the number of panels. In this example, it is 3.9 panels with spaces. Round this off to 4 panels plus spaces. Divide the total distance, 141 inches, by 4 to get a panel-plus-space width equal to 35¼ inches. The actual molding panel will be 32¼ inches, with a 3-inch open space. This is very close to the proportions of the Golden Rectangle.

Should you prefer shorter panels, select the length you want and use it in the calculations instead of the 33 inches used in **8–61**. There will be many times when the wall design will require the use of several different molding panel sizes, as shown in **8–62**. The goal is to get a balance of

8–62 Wall conditions frequently require new solutions to meet the special situation. As this is planned, try to keep the panels in a pleasing proportion.
Courtesy White River Hard Woods and The Hardwood Council.

pleasing, well-proportioned shapes that flow comfortably along the wall.

There are times when the molding panels on a wall are laid out in different sizes to provide

8–61 Mark the size of the molding wall panels by carefully measuring the length of the wall and making a layout to record the sizes calculated.

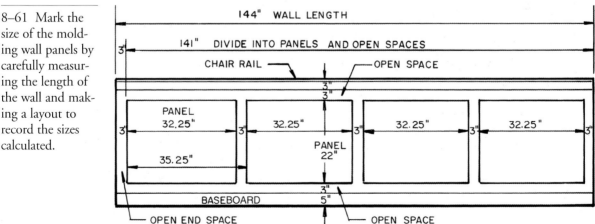

relief from the continuous repetition of the same size, as shown in **8–55** and **8–57**. Lay out various panel proportions on a drawing until you get a pleasing, balanced wall. In **8–59**, the molding panels were laid above and below the chair rail. Notice that the widths are the same in each section, providing vertical continuity.

ASSEMBLING THE PANEL MOLDING FRAME

The panel molding is cut to length with the corners mitered. It is then assembled, forming the frame. The miter joints can be glued and nailed in the same manner as those in picture frames. It helps to have some way to hold the corner closed as it is nailed. One way is to use a picture-frame clamp used by dealers that assemble and sell picture frames (**8–63**). The miter can be hand- or power-nailed. Nail from the side. Another way to hold the corner is to use a miter clamp (**8–64**). These clamps have needle-like points on the end that produce a tiny hole in the molding but are needed to hold the clamp in place.

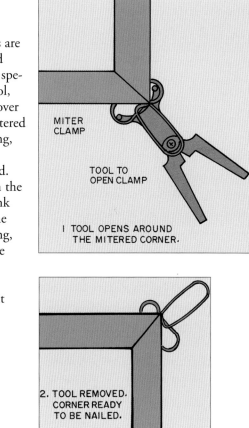

8–64 Miter clamps are opened with a special tool, fitted over the mitered molding, and released. Pins in the end sink into the molding, and the spring action holds it closed.

MITER CLAMP

TOOL TO OPEN CLAMP

I TOOL OPENS AROUND THE MITERED CORNER.

2. TOOL REMOVED. CORNER READY TO BE NAILED.

8–63 A picture-frame clamp can be used to hold the miter closed while it is glued and nailed.

INSTALLING THE MOLDING PANEL

Begin by running a chalk line or laser level beam along the wall, locating the top of the chair rail. Install the chair rail by nailing it to the studs. Then locate a line below it that is the guide for the top edge of the molding panel. This can be done with the chalk line or laser level, or by measuring down the width of the space between the chair rail and panel and running a line with a straightedge as shown in **8–65** and **8–66**. Now measure along the line, locating the sides of each molding wall panel (**8–67**). Extend the mark down some to give a clear side location (**8–68**).

Assemble the molding wall panel (**8–69**). Glue and nail each corner.

Apply a series of short beads of panel adhesive to the back of the molding, place the block of

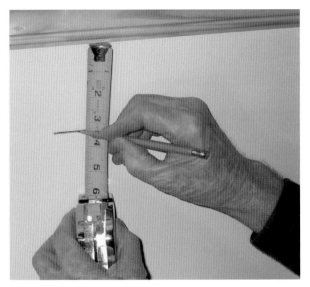

8–67 Mark the locations of the sides of each molding frame along the line.

8–65 Measure down the width of the space between the chair rail and top molding. Mark it in several places along the wall.

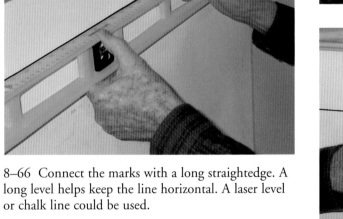

8–66 Connect the marks with a long straightedge. A long level helps keep the line horizontal. A laser level or chalk line could be used.

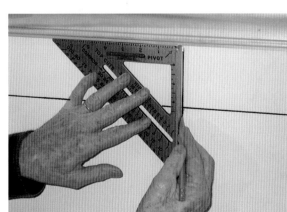

8–68 Mark the side locations with vertical lines.

wood against the chair rail, and butt the top of the molding panel to it (**8–70**). Press it against the wall surface. As you install it, check the side with a level to be certain it is plumb (**8–71**). Nail the top edge first so it hangs correctly. Then nail into the drywall on the sides and bottom. Power nailing produces the best results because it is faster and less likely to cause the panel to slip as the nails are driven. After the panels are installed, go back over them and caulk the inside and outside edges of the molding (**8–72**). The installation is now complete.

8–69 Cut and assemble the molding frame. The miters are glued and nailed. They can be held with a picture-frame clamp or supported with a block of wood cut accurately at 90 degrees.

8–70 Apply a series of short beads of glue along the back of the molding. Then place the top molding on the horizontal mark and align the sides with the vertical marks. It helps hold the molding for nailing if a wood block the same width of the space is placed between the chair rail and the top molding. The molding will not slip as it is nailed. Nail the top molding.

8–71 Then nail each side molding. Check it for plumb and adjust it as needed before nailing.

8–72 After the installation is complete and the glue has hardened, carefully caulk any slight cracks between the top and bottom edges of the molding and the wall. Also caulk any gaps in the miters in paint-grade molding and use special fillers designed for stain-grain molding.

CHAPTER 9

Finishing Interior Trim

As you consider how you might finish the trim, you'll notice more is involved than choosing the type of finish material available. The walls are the dominant feature of the room and how they are to be finished will influence the choice of finish on the trim. The walls are typically painted (**9–1**) or finished with wallpaper (**9–2**), while paneling is used less frequently (**9–3**). Wainscoting will influence the final appearance, as shown in **9–2** and **9–4**. The trim, walls, furniture, and other features of a finished interior should complement each other. This is accomplished by considering what materials, colors, and textures are used on them, and how visible they will be.

The trim finish is part of the overall color scheme for the room. Since the walls dominate,

9–2 In this room, the wall above the chair rail is finished with flowered wallpaper and the area below with a solid-color wallpaper. The trim has to be a color that is compatible with the wall covering. Courtesy Mr. and Mrs. James Sazama

9–1 The walls in this room were painted an off-white. The trim, stair skirt, and railing balusters were finished with a matching low-gloss enamel.

9–3 This paneled wall has matching trim on all sides, including baseboard and shoe molding.

they will usually receive the most consideration; the choice of trim finish should complement them.

The selection of color in a room greatly influences the atmosphere. For example, reds, yellows, and oranges are referred to as **warm colors**. Violet, blues, and greens are referred to as **cool colors**. The choice of the trim color must be in harmony with the wall color.

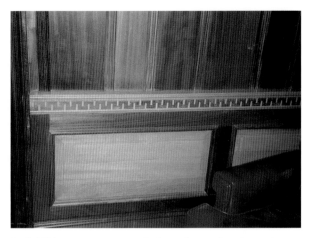

9–4 This room was paneled, and wood wainscoting was made using a different species of wood. Notice that the chair rail was made of a lighter color species to coordinate with the light wood panels.

While the trim could be the same color as the walls, it is often varied to provide a different look and give it some emphasis. It will have a different value or intensity. **Value** signifies the degree of lightness or darkness of the color. The trim could have the same color as the wall, but be of lighter or darker hue. This would give it some individual prominence, but not be overbearing. **Intensity** is used to describe the brightness and purity of the color. A color with strong intensity is rich and full. A color with low intensity has a flat look. By varying the intensity, the trim can be emphasized or downplayed.

Another factor in the choice of trim color is the architectural style of the house. For example, if your house is of the Early Colonial period, in which natural or stained wood trim has been used, that should be your guide. In the modern-style house with little architectural trim, the trim will be painted the same color as the walls, which is generally white or an off-white.

To sum up the situation, the choice is whether to blend the trim with the walls, develop a contrast, or vary the value and intensity to further enhance it.

While there are no strict guidelines to follow when choosing the color of the trim finish, the following schemes have been widely adopted. A **monochromatic color scheme** uses different values of the same color. The trim will be painted the same color as the walls, but slightly lighter or darker (**9–5**). This produces a slight but noticeable contrast. The trim softly frames the doors and windows, but does not dominate the room. The entire scene is one of close harmony.

9–5 The walls have a slight reddish-brown tint, which ties them in with the darker trim and sliding wood doors. **Courtesy Kolbe and Kolbe Millwork Co., Inc.**

Another scheme that has been widely used for many years involves **sharply contrasting colors on the trim and walls (9–2)**. Generally, the trim is a white or off-white. The value and intensity of the wall color can be varied to give the room the feel wanted, and the trim serves as a dominant architectural feature. The trim paint can have a gloss, semigloss, or flat finish.

A third color scheme has the **trim and walls the same color (9–1)**. It has no contrasting value or intensity; therefore, the trim blends relatively unobtrusively with the decor of the room. The trim does provide a clear, clean finish to doors, windows, and floor/wall and ceiling/wall intersections. It imparts a sense of order to the room.

Instead of painting the trim, some architectural styles traditionally require the trim be stained or have a natural finish (9–6). This is especially true when paneling is used and the trim is of the same species and finish as the paneling. However, stained trim can be used in the same manner as described for painted trim. The value and intensity of the stain used can be varied to give variety to the finished color. The application of a clear finish material over the stain can also affect the outcome. It can produce a gloss, semi-gloss, or natural wood. These are important considerations.

SOME PRELIMINARY CONSIDERATIONS

There is no shortcut to getting the trim ready to paint. A quality job requires careful preparation before the primer can be applied. While a quick once-over may get the job finished, over time defects and inadequate paint thickness will begin to appear. Since the trim is a major part of the appearance of a room and is expected to last a long time, high-quality materials and careful preparation are essential.

Another factor to consider is the way the finished job is left. There are no substitutes for neatness when applying paint and protecting walls and floors and for a careful, thorough cleanup. The use of drop cloths, sturdy brown wrapping paper, and masking tape are necessary, as are having available the solvent necessary to dissolve splattered paint and suitable rags and sponges. Clean constantly, not just at the end of the day.

LEAD-BASED PAINT

Lead-based paint was widely used until the mid-1970's, so if you are remodeling an older home, very likely the interior trim was finished with this type of paint. In 1971, Congress passed the Lead-Based Poisoning Prevention Act, and in 1976 the Consumer Product Safety Commission issued a ruling under the 1971 act basically banning the use of lead in paints used in residential buildings, furniture, and toys.

Those living in a house having trim finished with lead-based paint can be exposed to the lead by inhaling lead dust created as the old paint chalks, peels and chips off the trim. The dust set-

9–6 Staining the wood windows and doors, as well as the trim, makes them a dominant feature of the room. This installation features cherry wood. Courtesy Weather Shield Windows and Doors

tles on everything in the house and is ingested unknowingly. Ingestion of lead particles can have serious health consequences, including damage to the brain, kidneys, and red blood cells. It can cause partial loss of hearing, retard mental development and physical growth, and contribute to fetal injury.

CHECKING FOR LEAD

If you are remodeling an old house, you should first check to see if the trim has a coating of lead-based paint. Even if you do not plan to refinish the trim, you must check and, if it has lead-based paint, something should be done to remove it.

Consult the local building inspector to discuss the best way to check for lead-based paint. A portable X-ray fluorescence analyzer is typically used to check the level of lead in the paint. There are kits available at your local building-supply dealer. Examine what is available carefully to be certain it has been approved by the U.S. Consumer Product Safety Commission (**9–7**). A local laboratory or certified testing technician might be available to check sample paint chips removed from the trim. Your paint dealer may be able to make a recommendation as to where you can have tests run.

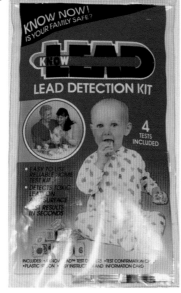

9–7 This is one of the lead detection kits available at building-supply dealers. Be certain the one you select is approved by the U.S. Consumer Product Safety Commission.

When you test for lead, remember that the trim in an old house has probably been repainted several times. The topcoat may be recent and lead-free, but coats below it could contain the old lead-based paint.

REMOVING LEAD-BASED PAINT

Removing lead-based paint requires considerable preparation, and professional advice and assistance should be considered. The Department of Housing and Urban Development, 451 Seventh Street SW, Washington, DC 20410, has bulletins providing information.

One solution is to remove the trim and install new material. This will create some dust and chips, and a high-quality respirator should be worn. Check your local codes to see exactly what provisions are required by law. A thorough cleanup afterward is required. The old trim and related debris are regarded as hazardous waste, so proper disposal is necessary.

Another approach is to remove the old paint. If removed by sanding, a high-efficiency particulate air vacuum filtration apparatus is required. Some chemical methods for removing the old paint can be used. If this is done, the surface must be absolutely clean. Any particles left will become dust when dry. A supplied air respirator and eye protection should be worn.

Basically, the solution is to not try to do this yourself. Employ certified people to handle the removal and disposal of the waste.

MOLDING SPECIES

When it comes to choosing a finish, the species of wood is considered. Some species are hard and have an open grain. Open-grain woods have small pore-like openings that must be filled as part of the finishing operation. Closed-grain woods are sealed with the paint primer or a clear coat. Some woods take paint better than others. Some woods, such as walnut, are so expensive it would be ridiculous to cover the beautiful grain and color with paint. A brief summary of the finishing characteristics is given in **Tables 9–1 and 9–2**.

The beauty of the natural woods used for interior trim is shown in **9–8**. The range of lightness and darkness from species to species gives the designer a broad pallet of natural color and beautiful grain to work with. Some of these woods, such as oak, are often stained to colors similar to those of other species.

PREPARING THE TRIM FOR PAINT

The goal is to have a smooth, defect-free surface to which the paint will adhere. With new moldings, this is not much of a problem. When refinishing old moldings, the problem is greater because the old finish may take some special treatment to get the paint to adhere to it and you must check the paint for lead.

One of the first things to do is to fill all nail holes, dents, and other damaged areas. If the trim carpenter did not set the nails, you will have to do it.

These repairs are best made with a latex or acrylic wood filler (**9–9**). They work easily, dry rapidly, and do not shrink. Fillers that shrink will leave a depression over each nail hole or dent. One way to fill a nail hole is to put a pad of filler on the end of a screwdriver and press it into the hole until it is full. You should leave a little extra filler above the hole so it can be

TABLE 9–1. SOFTWOOD FINISHING CHARACTERISTICS

Softwoods	Grain	Hardness	Paint-Holding Ability	Stain Quality
			1 = Best 5 = Poorest	1 = Best 2 = Good 3 = Fair
Fir, Douglas	Closed	Soft	4	3
Pine, Northern	Closed	Soft	4	1
Pine, Ponderosa	Closed	Soft	3	1
Pine, Sugar	Closed	Soft	2	1
Pine, Yellow	Closed	Soft	4	1
Red Cedar, Western	Closed	Soft	1	2
Redwood	Closed	Soft	1	2
Spruce, White	Closed	Soft	3	2

TABLE 9–2. HARDWOOD FINISHING CHARACTERISTICS

Hardwoods	Grain	Hardness	Paint-Holding Ability	Stain Quality
			1 = Best 5 = Poorest	1 = Best 2 = Good 3 = Fair
Ash, White	Open	Very Hard	3	1
Beech	Closed	Very Hard	4	2
Birch, Yellow	Closed	Hard	4	2
Butternut	Open	Medium	4	1
Cherry	Closed	Hard	4	1
Maple, Sugar	Closed	Very Hard	4	2
Oak, Red	Open	Hard	4	1
Oak, White	Open	Hard	4	1
Poplar	Closed	Medium	3	2
Walnut	Open	Hard	4	1

sanded flush with the surface of the molding. Let it dry overnight before sanding.

If screws were used to secure wide moldings, they will be countersunk and have a wood plug glued over the top (**9–10**). The hole for the screw is drilled, countersunk, and counterbored with a counterbore tool (**9–11**). If the molding will be painted, the plug can be of any similar wood. If it is a stain job, you can either make plugs from scrap pieces or, if a contrast is desired, use a lighter or darker plug (**9–12**).

To install the plug, wipe its sides with glue and put a little on the sides of the hole. Force the plug in place and wipe off any excess glue that squeezes out. You can press the plug in with your finger and then give it a light tap with a hammer (**9–13**). It should project about ¹⁄₁₆ inch above the surface of the molding and, after the glue dries, be cut flush with a belt sander. If more is sticking up than the sander can easily handle, lower the plug some with a sharp plane or chisel.

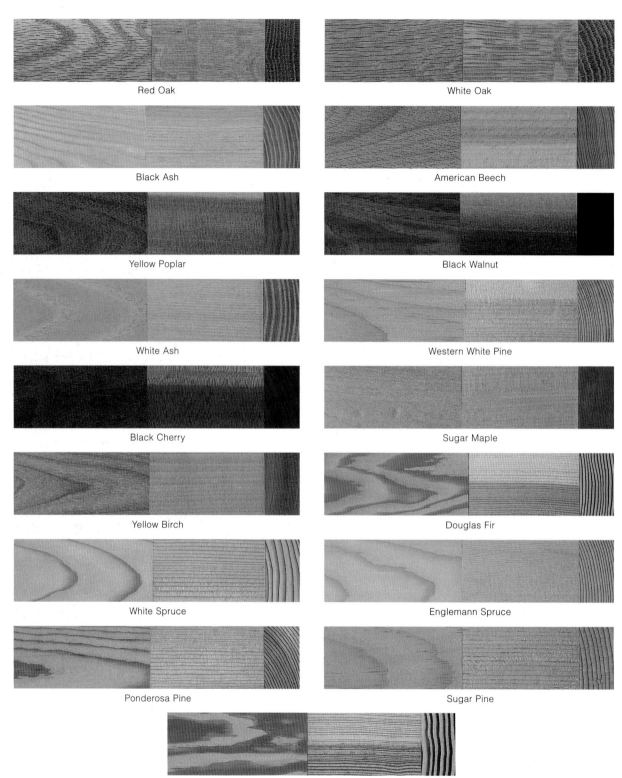

Red Oak

White Oak

Black Ash

American Beech

Yellow Poplar

Black Walnut

White Ash

Western White Pine

Black Cherry

Sugar Maple

Yellow Birch

Douglas Fir

White Spruce

Englemann Spruce

Ponderosa Pine

Sugar Pine

Shortleaf Pine

9–8 These samples show the range of lightness and darkness and the beauty of the grain of commonly used American woods. **Courtesy USDA Forest Service, Forest Products Laboratory.**

INTERIOR TRIM: MAKING, INSTALLING & FINISHING

9–9 Nail holes, cracks, and small defects can be filled with a latex or acrylic wood filler.

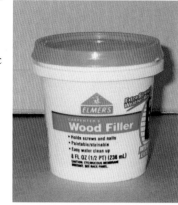

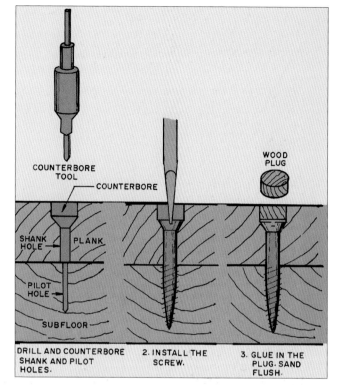

9–11 A counterbore tool.

9–10 Wood screws are countersunk and the holes covered with wood plugs. The holes are drilled, countersunk, and counterbored with a counterbore tool.

9–12 These plugs are of a wood of a different species than the molding, creating a decorative feature.

9–13 Put glue on the plug and lightly tap it into the counterbore. It should be left slightly above the surface of the molding so it can be sanded smooth.

Plugs are available at some building-supply dealers; however, you can cut them from scrap stock with a plug cutter (**9–14**). Install the plug cutter in a drill press and lower it into the wood. You will end up with a series of plugs that are cut in it. Remove them by cutting them off with a chisel (**9–15**).

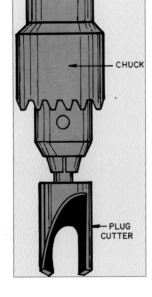

9–14 This plug cutter is mounted in a drill press and will cut plugs in a piece of scrap wood.

CHUCK

PLUG CUTTER

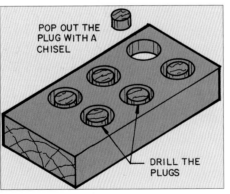

9–15 Drill the plugs in a piece of wood and pop them loose with a wood chisel.

POP OUT THE PLUG WITH A CHISEL

DRILL THE PLUGS

Dents can be repaired in paint-grade molding by spreading wood filler over it, letting it dry, and carefully sanding it flush.

Larger damaged areas are best repaired by cutting out the area and gluing in a wood patch. While this can be done with a wood chisel, a faster and neater installation can be made using a Lamello patch-cutting machine. This is shown in **Illus. 3–43 to 3–45** on page 67. Round patches can also be cut. After the patch has been glued in and dried, it can be planed down and sanded smooth.

PREPARING STAIN-GRADE MOLDINGS FOR A FINISH

The repairs on stain-grade moldings have to be the same color as the finished molding. Nail holes can be filled using **burn-in sticks**, which are usually made with a shellac or lacquer base. Burn-in sticks are available in a range of colors, so you have to select the one closest to the color of the finished molding.

The stick is melted with a **burn-in knife** (that can be bought at a hardware store or made), which is like a small soldering iron, but is heated with an open flame. It is heated and pressed against the stick over the hole or small defect. The melted material is pressed into the opening, and overflows the hole or defect slightly. After it has cooled, the excess is removed by sanding with a 280-grit abrasive paper wrapped around a sanding block.

Sometimes a wood dough product can be used. It is a putty-like material available in a variety of colors. It is pressed into the hole or defect with a putty knife. The defect should be completely full with a little excess above the surface of the molding. Keep it off the surrounding wood as much as possible, because if left on the surface, the wood around the defect will not accept stains. After it has dried, sand it smooth. If the defect is large or deep, consider applying several thin coats, allowing each to dry before the next coat is laid down.

Large patches of solid wood can be glued in place. You can machine patches from scrap wood to match the molding.

SANDING THE TRIM

Proper sanding of the molding is an important part of the finishing process. When possible, sand the strips of molding before they are cut to size and installed. It is also recommended the strips be primed front and back before cutting them to size.

The sanding papers available include garnet, a natural abrasive, and aluminum oxide and silicone carbide, which are man-made abrasives. The man-made abrasives last longer than garnet and, in general, leave a better finished surface. The grits most frequently required are 100, 150, 200 and 320. The larger the number, the finer the abrasive, so most of the time you will start with 100 grit and finish with one of the finer grits. Stain-grade moldings should be finished with a finer-grit abrasive paper than paint-grade because the stain will soak in even very tiny scratches and be quite noticeable, while paint will cover them up. Consider using a 220- or 280-grit paper.

Sheets are 9 × 11 inches and most finishers like to tear them into quarters to wrap around a sanding block (**9–16**), or cut them in half and fold them into thirds to get a narrower strip for hand sanding (**9–17**). Wood blocks can be used for sanding; however, rubber or felt blocks are flexible and more useful. The concave and convex surfaces can best be sanded with a thinner rubber or felt pad with the abrasive paper over it or by trimming pieces of foamed plastic insulation to fit the contour. Also available is a ¼-inch-thick rubber pad with an abrasive bonded on one side (**9–18**). It is very handy for curved and flat surfaces.

Sanding sponges are also good for sanding flat and curved surfaces. They have a flexible foam core to which an abrasive material is bonded (**9–19**). They are available with grits from 80 to 320.

Sometimes it will be necessary to use a power sander. Since they remove wood very rapidly, they are used sparingly. A **belt sander** (**9–20**) will cut very fast and can actually remove more of the surface than you planned, thus damaging the molding. They are used on flat surfaces that have heavy mill marks. A **finish sander** uses the same

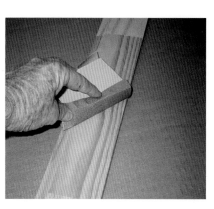

9–16 Rubber, felt, and foam plastic sanding blocks help when smoothing flat surfaces. The abrasive paper is wrapped around them.

9–17 For sanding in tight places, fold the quarter sheet of abrasive paper into thirds, making a stiff narrow pad. This helps sand in tight, narrow spaces.

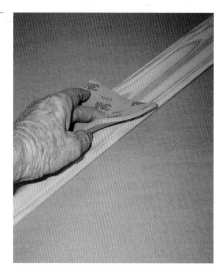

9–18 Thin rubber pads with abrasive bonded to the surface are very flexible and ideal for sanding curved and other irregular surfaces.

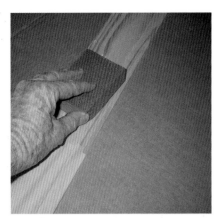

9–19 Sanding sponges work well with any type of finish and are flexible enough to use on larger curved surfaces.

9–21 A finishing sander will do most of the jobs where power sanding is necessary.

9–20 Belt sanders will remove a lot of wood in a hurry, so must be used with caution.

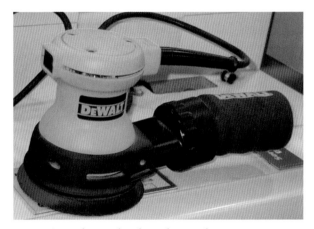

9–22 A random-orbital sander produces a very gentle sanding motion and is good for the final power sanding step.

abrasive sheets as mentioned earlier for hand sanding (**9–21**). It is often used after the belt sander and before final hand sanding. There are several types. The **random-orbital sander** (**9–22**) can be moved in any direction. **Straight-line sanders** must be moved with the grain. The vibrating and orbital sanding produces an across-the-grain pattern (**9–23**).

PREPARING OLD PAINTED TRIM

When repainting old trim, sand it to remove any gloss and smooth defects and chips. Use a medium-grit paper. This helps the new topcoat bond to the old finish. An alternative is to rub the trim with liquid sander-deglosser. This is a clear liquid that cleans and prepares the old surface to receive the new topcoat. It will cut the gloss so the new coat will bond.

Another technique employs paint remover. Applying it softens the old paint. When it appears wrinkled, scrape it off with a putty knife (**9–24**). Be careful not to scratch or damage the surface. Have lots of old newspaper to wipe the scrapings off the putty knife. Once all the paint has been scraped from the surface, rub it down with a stripping pad. A stripping pad looks somewhat like green steel wool; however, it will not splinter, rust, or shred like steel wool (**9–25**). Rinse the pad with solvent so it can be used again. Wipe any light residue left

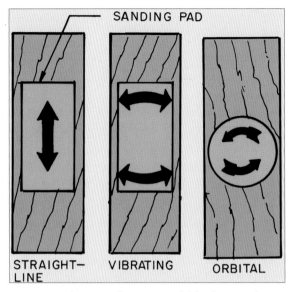

9–23 Finishing sanders are available that produce different sanding patterns.

9–24 A paste paint remover is easiest to use when paint has to be removed down to the bare wood.

on the surface with another dry stripping pad. Now wash the surface with the recommended solvent. Let the surface thoroughly dry before sanding and painting. Ventilate the room thoroughly and wear a respirator if recommended on the paint-remover can.

If the old trim was painted with an oil-based paint, it is very difficult to apply acrylic latex paint over it. The best thing to do is not try. Instead, apply oil-based paint over oil-based paint and acrylic latex paint over acrylic latex paint. If you attempt otherwise, the new coat often does not bond and peels away in large pieces.

If you must apply latex paint over oil-based paint, thoroughly sand the surface with 150-grit abrasive paper. Use a sanding sponge to get into corners and recesses. Thoroughly clean the surface so it is free of dust. A tack rag is a good thing to use. Then apply a bonding primer over the entire surface. Use one recommended by your paint dealer. This primer creates a fresh surface upon which the latex paint should bond. Both oil-based and acrylic latex primers are available.

Rub your hand over the old painted trim to see if it is sticky. Painted trim in kitchens and bathrooms often develops over the years a greasy, grimy surface. This must be removed by washing the trim with a good household cleaner. After rubbing it, thoroughly wash it with clean water and a sponge. Then sand the trim to smooth it and reduce the gloss.

If the old paint is peeling, it must be removed by scraping and sanding.

9–25 This green fibrous, nonmetallic stripping pad is used to remove any softened paint or varnish left after the excess has been removed by scraping.

MILL MARKS

Mill marks are sometimes found on the surface of moldings. They are a series of small parallel ridges left by the cutters of the machine shaping the molding. Some are very small and almost invisible. These should be sanded out or they will show as ripples after the molding has been finished. Larger, easily noticeable mill marks are difficult to remove and, if they are removed, the molding might show a shallow recess when it is finished. It is best to discard pieces with pronounced mill marks. Be certain to try to sand down mill marks before the molding is cut to length and installed.

TYPES OF PAINT

A visit to the paint dealer will reveal the array of paint products used on interior moldings. They also have samples so the color can be seen and coordinated with the wall color. It is a good idea to select the wall paint, wallpaper, and trim paint from the same paint store. Close comparisons can be made and the store's salesperson can offer valuable suggestions.

Trim should be painted with a semigloss or gloss paint. The gloss coating is harder than the flat coating and will withstand some bumps and cleaning. However, it has a shine that many people do not like.

The paint dealer can also mix colors to your requirements. After a preliminary mix has been made, paint a piece of scrap stock and let it dry. (The color is often quite different after it is dry.) While a professional painter can use a white paint base and add color to get the hue desired, it is better to let the paint dealer do it. They mix according to a formula, which they record in their records, so that it is easy to get more paint of the same exact color years later.

Do not buy the cheapest paint available. The total cost for the paint is so small it is best to purchase the highest quality available. One way to tell if the more expensive paint is really better is to check the percentage of volume solids. **Volume solids** include the pigment and binder that are suspended in the solvent. The higher the percentage of solids, the better the paint. Percentages of 35 to 45 are in premium-quality paints. Paints using **titanium dioxide** as the pigment are superior.

Also examine the information on extenders. **Extenders** are low hiding-power pigments added to paint to increase coverage. There are inexpensive additives such as clay, calcium carbonate, and silica. Extenders compromise the durability of the paint and reduce the chances for one-coat coverage. The less extender added, the better.

The paint dealer will also get you the recommended **primer** for the type of paint selected. This is very important.

LATEX PAINTS

Latex paints have a water vehicle into which latex emulsions, pigments, and binders are suspended. The binders may be 100 percent acrylic, styrene acrylic, or vinyl acrylic. These affect the gloss, stain resistance, adhesion, and color retention. Latex paints with **100 percent acrylic binders** are especially durable and are found in high-quality paints.

Latex paints are affected by the surface upon which they are applied. This is especially true for semigloss and gloss latex enamels. The surface upon which they are applied must be absolutely free of dirt, soap, and grease. This is especially a problem if you are repainting an older kitchen or bathroom. Improper preparation will cause the latex enamel to crawl or sag on the surface. Some latex paints are formulated with glycols and water as part of the solvent. Since glycols are water-soluble, they release water slowly and the enamels may run or sag, especially if the humidity in the air is high.

Water is the solvent for cleaning brushes and doing other clean-up jobs.

Latex primer is a water-thinned paint used as a first coat on all unpainted moldings. Some people apply a coat for regular latex paint as the first coat instead of a separate primer.

ALKYD PAINTS

Alkyd paints use an alkyd resin, which is a type of thermosetting plastic. They are oil-based and thinned with a solvent such as turpentine and can be cleaned up with paint thinner or mineral spirits. They have a strong odor that may last several days. Alkyd paints are more resistant to damage from frequent cleaning than latex paints. They are also not as sensitive to the surface preparation as latex paints. They are widely used in kitchens and bathrooms where humidity and frequent cleaning is necessary.

Use the primer made by the company that manufactures the alkyd paint.

APPLYING THE PRIMER

After the trim has been sanded, the next step is to put primer on both sides. The **primer coat** fills any small irregularities in the surface and provides the base upon which the topcoat will bond. After it is thoroughly dry, lightly sand it to remove any areas that are still a bit rough and to get as level a surface as possible for the topcoat. Primers for oil-based paints are sometimes a bit thick in the can. Thin it to a cream-like consistency with paint thinner. Latex primers are water-based and can be thinned with water if necessary. Your paint dealer may have a primer that can be used for both oil- and latex-based paints.

As you sand the primer coat, do it very lightly so you do not cut back and expose bare wood. If bare spots appear because the molding required considerable sanding, recoat the bare spots or, better still, apply a complete second coat of primer. You could mix a little of the top-coat enamel in the second coat of primer if you wish. Remember, if possible prime both sides of the long pieces of molding before they are cut to size and installed. The primer can be sprayed on, greatly speeding up the process and producing a quality, uniform thickness coating.

CAULKING THE TRIM

Any minor cracks between the molding and the wall or in mitered or coped joints should be caulked with a flexible caulk. An acrylic latex caulk is widely used. Be certain it is paintable. Paint will not bond to some types of caulk, and this could create a mess. Caulking is done after the trim has been primed, nail holes filled, and dents repaired. It provides a tight joint between the meeting materials. Some people caulk the base and casing even if there are no visible cracks, so there is a smooth transition between them and the wall.

Generally, the caulk is applied with a caulking gun that holds a tube of caulk (**9–26**). There are a number of types of caulking guns available. It is important to select one that permits the rapid release of pressure on the caulk so that when you finish a run, the caulk does not keep pouring out. Caulk is also available in small tubes much like toothpaste (**9–27**). These are easy to control

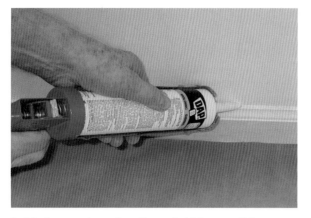

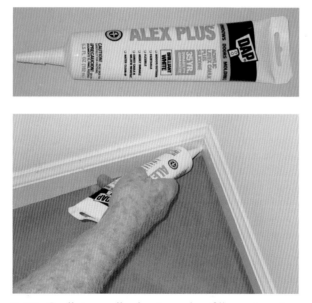

9–26 Large tubes of caulk are held in a caulking gun, which forces the caulk out the nozzle.

9–27 Caulk in small tubes is used to fill openings in tight, difficult places.

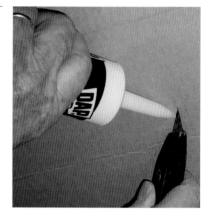

9–28 Open the tube of caulk by cutting the end of the nozzle on an angle. The more of the nozzle removed, the larger the size of the caulk bead extruded.

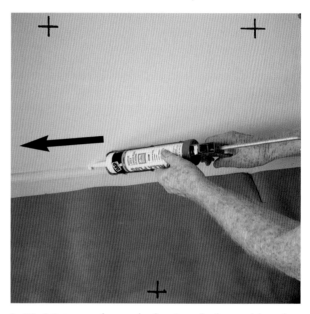

9–29 Most people get the best results by pushing the nozzle along the edge to be caulked.

and are especially useful when caulking a short run such as a mitered corner.

To use the caulking gun, insert the cartridge in the gun. Then cut the plastic nozzle at a sharp angle (**9–28**). When in use, the cut surface should be able to fit into a corner formed by the trim and the wall. In most cases, it works best to push the caulking gun down the trim because this tends to force the caulk into the crack (**9–29**). After laying a bead of caulk, run a wet finger over the joint (**9–30**). Clean excess caulk off the trim with a wet cloth. In corners, miters, and coped joints, clean the caulk off the trim so that the corner is clean and sharp and only the crack is closed.

When you are finished caulking, clean the end of the nozzle with a wet cloth and screw a large electrical connector over the end. This seals the cartridge so that the caulk in the nozzle does not harden (**9–31**).

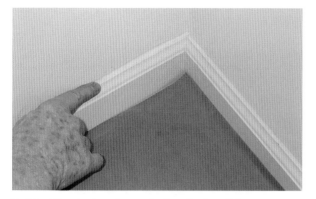

9–30 After laying down the bead of caulking, you can force it into the opening and smooth it by wiping it with a wet finger.

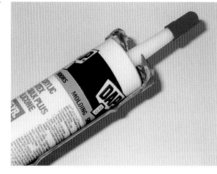

9–31 Seal the nozzle with an electrical connector to keep the caulk from hardening in the nozzle.

9–32 The brush on top is a sash brush. The end of the bristles is slanted. The lower brush is a standard brush.

BRUSHES

It takes some practice to apply the enamel topcoat with a brush. **First, use a China boar-bristle brush with oil-based paints.** They are more expensive than synthetic bristle brushes, but do a better job with oil-based enamels and, if properly cleaned and stored, will last many years. Do not use China bristle brushes with latex paints. Latex paints have a water base that will damage the bristles on a China bristle brush.

A nylon or polyester bristle brush should be used with latex paints.

For most trim work, you can use a 2-inch **sash brush** that has a diagonal tip, which helps when painting near the wall, and a 2 or 2½-inch **flat brush** with a square edge for finishing the flat surfaces and wide, sculpted areas (**9–32**). If the molding is very wide, a 3-inch flat brush will save some time.

When buying brushes, remember that cheap brushes do not hold the quantity of paint or produce as fine a finished surface as the more expensive brushes.

Brushes should be cleaned in the solvent indicated on the paint can. To do this, pour the solvent into a container and work the brush into it, squeezing the solvent out of the bristles with your finger. Throw the dirty solvent away and clean the brush again in fresh solvent. Usually three cleanings will do the job.

Now, comb the bristles so they are straight. Hang the brush until the bristles are dry and then wrap them in paper (**9–33**). Store them flat or by hanging the brush from the hole in its handle.

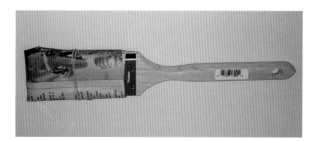

9–33 After the bristles are clean, brushed straight, and have dried, wrap them in paper and store the brush by hanging it by the handle or by letting it lay perfectly flat.

APPLYING THE TOPCOAT

Generally, most new painted trim will require two topcoats over the primer. Some very high-quality enamels will cover with one coat. When refinishing old trim that has a dark finish, two coats are almost always necessary. If the old finish was chipped or damaged, extensive sanding will be required. Some people only paint over the damaged areas and, when they are dry, apply the first topcoat on all the trim.

Remember, the trim should have a gloss or semigloss topcoat. These topcoats are harder, but hold up better when cleaned.

After the prime coat has been sanded, wipe off any sanding dust. Prepare the topcoat by mixing it thoroughly. Some people use a metal paddle in an electric drill, while others use a wood stirring stick, available at no cost from the paint dealer. If you plan to use a high-quality enamel, one coat will often be enough. If you prefer a two-coat enamel finish, it should be noted that some people prefer to thin the enamel about 10 to 15 percent by volume. The brush should flow smoothly along the molding. If it tends to stick or is a bit difficult to move across the surface, thin the enamel a little more.

Do not paint out of the full can of paint. Stir the paint in the original can. Then pour some of the paint into a paint bucket and paint from it. (**9–34**). Pour about a quart into the bucket, thin it as necessary, and paint from the bucket. A small paint tray is shown in **9–35**. It is convenient to handle and can also be used to clean brushes. When using a paint bucket or tray, you can close up the can of paint so that the solvent does not evaporate and cause the paint to thicken.

To load a brush, dip it into the paint about two inches, raise it up, and either gently draw each side of the brush against the wire in the bucket (**9–36**) or lay the brush against the side of the bucket and tap each side of the brush (**9–37**). This leaves enough paint in the brush to

9–34 Stir the paint in the original can and then pour some into a paint bucket and paint from it.

9–35 A small paint tray holds a limited amount of paint and is very useful when you are working in areas where a heavier paint bucket may pose a problem.

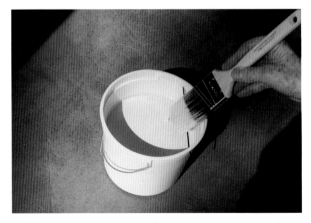

9–36 Dip the brush into the paint about two inches and lightly smooth each side of the bristles against the wire on the bucket.

9–37 After dipping the brush about two inches into the paint, you can wipe off the excess by running the bristles gently over the side of the paint bucket.

9–38 A skilled painter can cut the paint next to another surface with a sash brush.

lay down and smooth out. When you are covering wide surfaces, less paint needs to be wiped out. Just tap out enough so that it does not drip off the brush as you move the brush from the bucket or tray to the work.

While there is no standard way to start painting trim, the following suggestions are typical. Frequently, the walls will already have been painted, so the edge of the trim next to the wall will have to be handled carefully. Some people are skilled enough to cut in the paint freehand, keeping it off the wall (**9–38**). Others will use a metal shield, which is placed next to the strip along the wall (**9–39**). However, many people prefer to paint the trim before the walls. This can save a lot of time.

Remember to cover the floor with drop cloths. If the floor has been carpeted, extra care is necessary because you cannot wipe off small drops as you can with wood or vinyl flooring.

When painting the baseboard, chair rail, or crown molding, begin in a corner (**9–40**). Do not stop until you reach the next corner. This will help avoid an overlap between coats that will show when the job is finished. Begin by painting the trim on the top and bottom edges for three or four feet and then fill in the field (**9–41**). Then do another section until the opposite wall is reached.

9–39 A metal shield can be used to protect the adjoining surface. Be certain to clean the shield frequently.

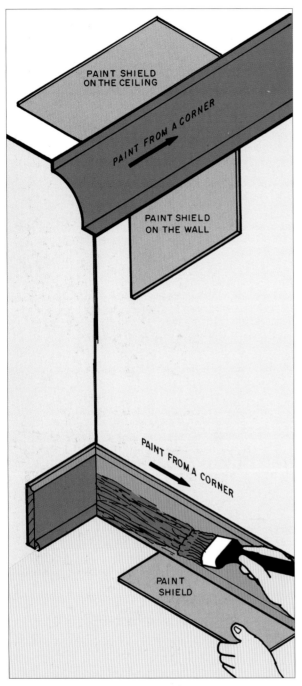

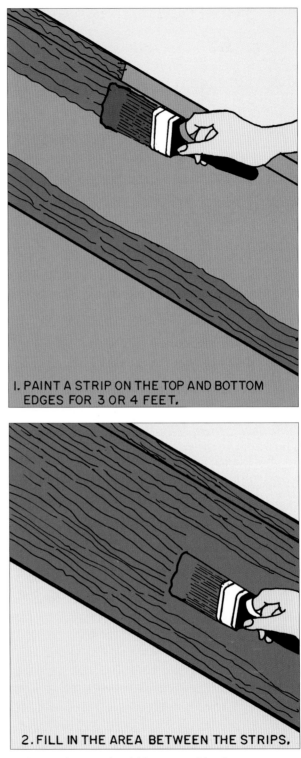

1. PAINT A STRIP ON THE TOP AND BOTTOM EDGES FOR 3 OR 4 FEET.

2. FILL IN THE AREA BETWEEN THE STRIPS.

9–40 When painting baseboard, chair rails, or crown molding, begin in a corner and paint to the next corner. Never leave a section only partially painted.

9–41 Wide trim should be painted by first coating the top and bottom edges for three or four feet and then filling in the area between them. Then do another section.

PAINTING WINDOWS

The procedure for painting windows using double-hung sash is illustrated in **9–42** and **9–43**. While others may prefer a different sequence, this is typical. Do the following:

1. Mask the glass with masking tape to keep the paint from smearing it. If masking tape is not used, as you work constantly wipe off any excess paint before it dries. It can be scraped off with a razor blade later, but this is messy and time-consuming.

2. Lower the top sash about an inch and paint the exposed stiles, rails, and muntins (**9–42**).

3. Raise the lower sash almost to the top of the frame. Lower the top sash almost to the sill (**9–43**). Paint the unfinished bottom end of the upper sash. Paint the entire upper sash.

4. Let the sash dry before painting the window casing, sill and apron. Move the window several times during the drying time so it does not stick to the parting strips or the metal channel in which it slides.

5. If the windows are very old, they will have wood parting strips that form channels into which they slide. Paint these after the sash has been painted and dried (**9–44**).

The order to follow when painting the sash varies. Each painter has a procedure they prefer. Some people prefer to paint the stiles and rails and then the muntins. Others keep to the following order, painting the horizontal muntins, the vertical muntins, the top and bottom rails, and finally the stiles. Whatever you do, establish a procedure and follow it.

To paint the window casing, paint the top casing first completely across the window. Then, starting at the top, paint each side casing down to the sill. Some people prefer to paint the sill and apron last (**9–45**).

There are other types of windows available. Apply these instructions to them, also.

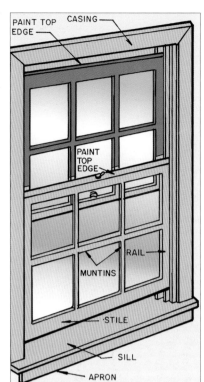

9–42 Begin painting a double-hung window by lowering the upper sash an inch or so and painting the exposed surfaces of the stiles, rails, and muntins. Move the sash occasionally so it does not stick.

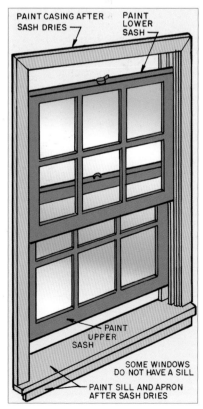

9–43 After painting the top area of the upper sash, raise the lower sash and slide the upper sash down. Paint the rest of the upper sash and the entire lower sash. Move the sash occasionally so they do not stick.

9–44 If the windows have wood parting strips, paint them after the sash have dried. Begin by lowering both sash to the sill and painting the top of the parting strips. When these are dry, raise the sash and paint the strips on the lower end. Keep paint off the sides and inside the track. Rub paraffin or wax on the sides of the parting strips.

9–45 After the sash have been painted and have dried, paint the casings, sill, and stool.

PAINTING DOORS

To paint a door, first remove as much hardware as is practical or cover it with masking tape. This includes covering the hinges. Hinges with paint smeared on them make a bad impression. Some people prefer to take the door off the hinges and lay it across padded sawhorses for easy application; however, this means you only paint one side at a time. Be certain all the edges are sealed, including the top and bottom edges, so moisture cannot enter them, possibly causing the door to swell or twist.

Wood doors are sanded and primed as discussed for preparing the trim. Fiberglass doors must be finished as recommended by the manufacturer. The local paint dealer will show you products designed for this job. For example, there is a stain containing petroleum distillates that can be used on wood or fiberglass. The manufacturer supplies a polyurethane clear finishing topcoat (**9–46**). Two coats are recommended.

Fiberglass doors can also be finished with the same paints used on wood doors. First, sand the old clear coat to produce a bonding surface.

9–46 This fiberglass door was colored with a manufacturer's recommended gel stain and two coats of polyurethane.

Then paint it with the same type of primer used for a wood door. Finally, coat with a latex or alkyd enamel.

One way to paint **paneled doors** is detailed in **9–47**. Do the following:

1. Seal the top and bottom edges of the door.
2. Paint the molding around the edge of the panels and then the panels. Complete each panel in the order shown in **9–47** (steps 1, 2, 3 and 4). Do not let the paint overlap on the faces of the rails and stiles. If it does, wipe it off. Start at the top of the panel and paint to the bottom.
3. Paint the top, lock, and bottom rails (steps 5, 6 and 7).

4. Paint the stiles, beginning at the top of the door and working to the bottom (steps 8 and 9).

If the door is new, prime the top and bottom edges to seal out moisture. Do not work up to a thick paint layer; just a sealer coat is fine.

If the door has glass panes, first mask the glass. Then paint the horizontal mullions and then the vertical muntins on one pane (**9–48**). Move on to the next pane and repeat these steps.

When painting **flush doors**, you must work quickly because you have a large area to cover and it is important to avoid overlaps of each strip painted. You may want to thin the paint a little to help make it easier to spread, but not enough to reduce coverage.

9–47 Paint the doors in a logical order by completing the panels first and then the stiles and rails. Start at the top and work toward the bottom of the door.

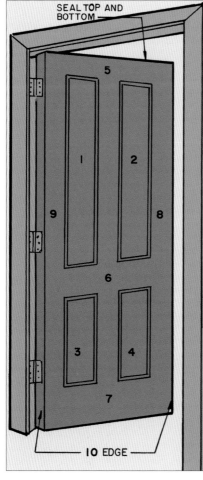

1. PAINT THE HORIZONTAL MULLIONS.
2. THEN PAINT THE VERTICAL MUNTINS. MOVE TO THE NEXT GLASS AND REPEAT THESE STEPS.

9–48 If the door has glass panes, paint the horizontal muntins first and then the vertical muntins.

Following is a suggested plan for painting flush doors (**9–49**):

1. Begin in an upper corner and lay on vertical strokes about one-third to one-half of the way across the door and about one-fourth the length.
2. Then stroke this area horizontally without putting more paint on the brush.
3. Repeat this until you have painted across the top of the door.
4. Now paint another strip across the door. Brush from the above coat down over the area below. You are painting from a surface beginning to dry onto the freshly painted below.
5. Repeat until the entire door is covered.
6. Paint the edges of the door the same color as the door.

FINISHING STAIN-GRADE MOLDING

After the stain-grain molding has been sanded, use a soft brush to clean the dust out of the pores of the wood. Some woods have an open grain, which means they have pore-like openings. Species like this often used for interior trim include oak, walnut, mahogany, and ash. These require the pores be filled with a paste wood filler (see page 197). Other stain-grain woods, such as maple, birch and cherry, have a closed grain and do not require a paste filler.

Before any finishing begins, the decision must be made whether to paint the walls and then the trim, or do the trim first. Many prefer to do the trim first.

CLEAN-UP

Next, clean the room thoroughly. Vacuum the floor, even if it is just the rough subfloor. Wipe down the trim with a tack rag. A tack rag is a cloth that has a sticky surface. The dust sticks to the rag rather than being blown into the air, only to resettle on the trim. If a rough subfloor is exposed, consider mopping it and keeping it damp to hold down any dust that may still be present. Now apply the required sealers, stains, fillers, and topcoats.

9–49 Paint flush doors in horizontal strips. Paint in sections so the paint will be smooth and flow together across the door. Paint from the unpainted area toward the just-painted area.

1. PAINT A SECTION STARTING IN AN UPPER CORNER CORNER USING VERTICAL STROKES. THEN BRUSH WITH HORIZONTAL STROKES.

2. THEN PAINT THE NEXT AREA ACROSS THE TOP.

3. PAINT THE NEXT STRIP. PAINT INTO THE PREVIOUSLY PAINTED AREA.

4. CONTINUE DOWN THE DOOR IN HORIZONTAL STRIPS.

It is also important to keep the topcoat free of impurities. Put only a small amount in a paint bucket (**9–34**) or trim pan (**9–35**). In this way, the entire can will not be open to dust and begin to set up. After you have been working several hours, the can will have been opened several times, so strain the clean topcoat in it and in the bucket through a paint strainer (**9–50**) or an old nylon stocking to remove lumps and other particles. Strainers are available wherever paints are sold.

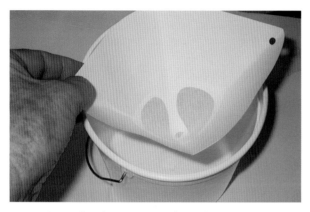

9–50 Stain the clear topcoat through a cone-shaped stainer every now and then to remove any lumps and impurities that may have gotten into it as it was being used.

STAINS

Stains are applied to the wood to change the color from its natural tone to a darker one or to add a bit of color, such as a hint of red. They are also used to make a less costly wood look like a more-expensive species. Sometimes butted moldings of the same species have quite different color variations, and careful staining can bring them together so that the colors in both appear the same.

Two types of oil stains are generally used: pigmented and penetrating. **Pigmented oil stains** are available in a wide range of colors. They are applied by wiping them on with a cloth pad or brushing. They lay on top of the sealer coat, so do not penetrate into the wood. After they have been in place for five to ten minutes, carefully wipe off the excess with a clean, soft cloth. Wipe across the grain to be certain coverage is complete and remove any excess that has accumulated in corners and ridges. The longer the stain remains on the surface, the darker it gets. Try some on scrap molding and develop the drying time to be used. Allow it to dry 24 hours before sealing. Temperature and humidity can change the drying time.

Penetrating oil stains are available premixed in a range of colors. They are a combination of oil-soluble dye with a vehicle such as turpentine, naphtha, or benzol; These stains penetrate the wood. After brushing them on, let the molding dry for the time recommended on the can. Then wipe off the excess. Good timing is important to get a uniform tone on all the trim in the room. These stains usually dry in 12 to 24 hours.

PASTE WOOD FILLERS

Paste wood fillers are used to fill the pores in open-grain woods. They contain a **filler** material, such as silica, or some other fine, inert material, a **vehicle** consisting of boiled linseed oil, turpentine or benzine, sometimes a **coloring agent**, and a **binder**. These are mixed to form a paste. They are sold in neutral tones and a variety of colors to match the various species of wood.

The fillers are spread on the wood with a brush (**9–51**) and worked into the pores. First brush it on with the grain and then brush across the grain. Some people do this with a stiff bristle brush or rub it with a cloth. Wear rubber gloves because it will stain your hands.

Allow the filler to set a few minutes. When it loses its glossy appearance, wipe it off across the grain with a heavy cloth (**9–52**). If you wipe with the grain, you will remove much of the filler from the pores. Wipe until the surface is free of filler and give a final gentle wipe with a clean cloth with the grain. Filler will get into corners of the molding. Carefully remove it with a small pointed stick.

9–51 Brush on the filler first with the grain and then brush it across the grain. Force it into the pores of the wood.

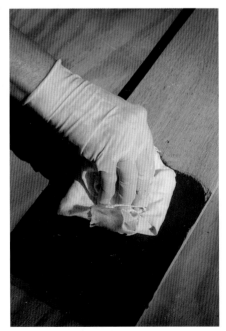

9–52 After the filler has set the required time, wipe off the excess across the grain, leaving the pores filled.

VARNISHES

The topcoat over the filled and stained molding is usually some type of varnish. **Natural varnishes** are one of the oldest finishes used for protecting wood surfaces. At the local paint store, you will find there are a variety of types available and that the staff can give you information about the advantages and disadvantages of each. For example, some dry so fast that it is difficult to get smooth coverage using a brush unless you work rapidly. Others dry very slowly, so if there is any dust in the room it could settle and stick to the varnish.

Linseed oil varnish is the older, traditional varnish. It is available in three varieties: short-oil, medium-oil, and long-oil. For interior trim work, medium-oil varnish is the type generally used. Another type of varnish is the **alkyd-resin** type, which uses a chemically altered linseed oil. It dries in about four hours and provides a fair, scuff-resistant finish. A third type is **polyurethane**, which has plastic chemical ingredients. It dries in two to three hours and produces a hard, durable, scuff-resistant finish. It is probably the most widely used topcoat on interior trim. Special sealers for polyurethane finishes are available. Use the sealer made by the company that made the polyurethane (**9–53**).

Varnishes are usually applied by brushing. If applied by spraying, considerable masking of the walls and nearby surfaces is necessary.

9–53 Polyurethane is a clear topcoat that is often used on interior trim.

STAINING OPEN-GRAIN WOODS

Following is a plan for staining and finishing open-grain woods:

1. Sand and clean the surface with a stiff brush. Wipe it with a solvent wood cleaner on a cloth. Then wipe off any residue with a cloth dampened with denatured alcohol.
2. If using a pigmented oil stain, apply a sealer recommended by the manufacturer. If using a penetrating oil stain, apply a special sealer made for penetrating stains. Some prefer to **fill the grain before staining.**
3. Seal the wood with a thin coat of shellac or a sanding sealer. Let it dry.
4. Apply the paste filler that has been colored to match the stained wood. Test it on a scrap piece before putting it on the molding. Allow it to dry for 24 hours. If you choose to omit this step to save the labor of filling and apply an additional coat of varnish, the open pores will still be somewhat visible, but some find this acceptable. Also, some prefer to fill the grain before staining.
5. Apply a sanding sealer. After it is dry, sand it lightly with 220-grit paper. Do not cut through into the stain. The sealer prevents the stain from bleeding into the clear topcoat.
6. Apply the first coat of clear topcoat. Let it dry. Sand it lightly to reduce the gloss. Be absolutely certain the coat is dry before trying to sand it; otherwise, you will have a rubbery mess.
7. Apply the second clear topcoat. Allow it to dry. Add additional coats if necessary.
8. Rub the last coat with 400-grit wet or dry abrasive paper, using water as the lubricant. Rub carefully so you do not cut through the finish. Sand just enough to smooth out the coating and adjust the gloss.

STAINING CLOSED-GRAIN WOODS

Following is a plan for staining and finishing closed-grain hard and softwoods:

1. Prepare the surface and brush it clean. Wipe it with a solvent wood cleaner to remove dirt, grease, and finger marks. Then wipe it down with a cloth dampened with denatured alcohol.
2. If using a pigmented oil stain, apply a sealer recommended by the manufacturer, or 3-lb.-cut shellac. Let it dry and fine-sand it. If using a penetrating stain, apply a special stain-control coating that is designed especially for use under penetrating stains.
3. Apply the penetrating or pigmented stain as directed by the manufacturer.
4. Apply a sanding sealer coat. Use the sealer recommended on the stain can. Dry-sand the wood lightly with 220-grit abrasive paper. The sealer prevents the stain from bleeding through the topcoat.
5. Apply the first clear topcoat. Let it dry. Then sand it lightly with 220-grit abrasive paper. Be absolutely certain it is dry before trying to sand it; otherwise, you will have a rubbery mess.
6. Apply the second topcoat. Let it dry. Sand it with 220-grit abrasive paper if a third coat is to be applied. If not, lightly sand it with 400-grit wet and dry abrasive paper using water as the lubricant.

TOUCHING-UP STAINED TRIM

Trim will, eventually, get a few nicks and scratches. If it is painted, you can sand the area around the scratch and apply a coat of paint. Feather out the old paint around the scratch so the new coat will flow smoothly over the damaged area.

Stained trim can have minor scratches and nicks touched up with a marker containing a stain (**9–54**). The stain marker is available in a range of colors, so you can select one as close to the trim color as possible. Apply the stain by brushing it on with the tip (**9–55**). Carefully wipe off the excess stain with a soft cloth.

Another way to repair scratches and cover nail holes is to use a crayon-like material in the form of a pencil (**9–54**). They are available in a range of colors. Work the material into the scratch or hole by rubbing the pencil across it (**9–56**). Then carefully wipe off the excess material and smooth the material in the damaged area.

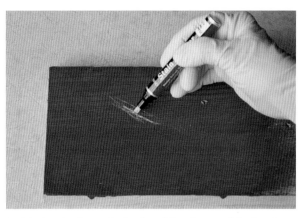

9–55 Apply the stain by brushing it on with the tip of the marker. Wipe off any excess with a soft cloth.

9–54 These products are used to repair small nicks, scratches, and holes in stained interior trim.

9–56 Rub the wax-like core of the pencil into the scratch or small hole. Wipe off any excess with a cloth and smooth the material in the defect.

Index

Metric Equivalents

[to the nearest mm, 0.1cm, or 0.01m]

inches	mm	cm	inches	mm	cm	inches	mm	cm
⅛	3	0.3	13	330	33.0	38	965	96.5
¼	6	0.6	14	356	35.6	39	991	99.1
⅜	10	1.0	15	381	38.1	40	1016	101.6
½	13	1.3	16	406	40.6	41	1041	104.1
⅝	16	1.6	17	432	43.2	42	1067	106.7
¾	19	1.9	18	457	45.7	43	1092	109.2
⅞	22	2.2	19	483	48.3	44	1118	111.8
1	25	2.5	20	508	50.8	45	1143	114.3
1¼	32	3.2	21	533	53.3	46	1168	116.8
1½	38	3.8	22	559	55.9	47	1194	119.4
1¾	44	4.4	23	584	58.4	48	1219	121.9
2	51	5.1	24	610	61.0	49	1245	124.5
2½	64	6.4	25	635	63.5	50	1270	127.0
3	76	7.6	26	660	66.0			
3½	89	8.9	27	686	68.6	inches	feet	m
4	102	10.2	28	711	71.1			
4½	114	11.4	29	737	73.7	12	1	0.31
5	127	12.7	30	762	76.2	24	2	0.61
6	152	15.2	31	787	78.7	36	3	0.91
7	178	17.8	32	813	81.3	48	4	1.22
8	203	20.3	33	838	83.8	60	5	1.52
9	229	22.9	34	864	86.4	72	6	1.83
10	254	25.4	35	889	88.9	84	7	2.13
11	279	27.9	36	914	91.4	96	8	2.44
12	305	30.5	37	940	94.0	108	9	2.74

Conversion Factors

1 mm	=	0.039 inch	1 inch	=	25.4 mm	mm	=	millimeter
1 m	=	3.28 feet	1 foot	=	304.8 mm	cm	=	centimeter
1 m²	=	10.8 square feet	1 square foot	=	0.09 m²	m	=	meter
						m²	=	square meter